P9-ARA-449

BIN●CULAR STARGAZING

Mike D. Reynolds

STACKPOLE
BOOKS

0   11557 03136   2

3 1267 13919 8343

Copyright © 2005 by Stackpole Books

Published by
STACKPOLE BOOKS
5067 Ritter Rd.
Mechanicsburg, PA 17055
www.stackpolebooks.com

All rights reserved, including the right to reproduce this book or portions thereof in any form or by any means, electronic or mechanical, including photocopying, recording, or by any information storage and retrieval system, without permission in writing from the publisher. All inquiries should be addressed to Stackpole Books, 5067 Ritter Road, Mechanicsburg, Pennsylvania 17055.

Printed in the United States of America

10 9 8 7 6 5 4 3 2 1

First edition

Cover design by Wendy Reynolds
Cover photograph of Tarantula nebula by Corbis
Art on page iv by Ted Cox
All photos by the author except where otherwise indicated

Library of Congress Cataloguing-in-Publication Data

Reynolds, Michael D.
    Binocular stargazing / Mike D. Reynolds.
        p. cm.
    Includes bibliographical references.
    ISBN-13: 978-0-8117-3136-2
    ISBN-10: 0-8117-3136-7
    1. Astronomy—Observers' manuals. 2. Binoculars. I. Title.

QB63.R49 2005
522—dc22

2005025461

Contents

Acknowledgments

Simply put, I love binocular astronomy. I have a number of telescopes (as my bride will correctly tell you, a few too many), yet I always carry a pair of binoculars with me regardless of my observing plans. But this was not always the case. Years ago, my then new wife, Debbie, and I would take our telescope out observing, and soon she began insisting on my bringing binoculars. I had a pair of old 7×35s I had been using since I was a kid, which are now more than forty years old. She'd use the 7×35s and I'd use the telescope. It seemed as though I would go over and use the binoculars when she "discovered" some jewel in the sky more often than she'd come over to the telescope.

The reason was simple: The low magnification, wide field of view, and ability to use both eyes allowed me to explore—to take in—the heavens on a different and, in some ways, more compelling level. Throughout my days as an educator and museum professional, people often asked me which telescope they should buy. My answer always surprised them.

So I guess this book has a long history, since I first lifted those inexpensive 7×35s skyward more than forty years ago. *Binocular Stargazing* is built on that history, as well as the assistance of a number of people. Peter Abrahams, fellow Antique Telescope Society member, has done a significant amount of historical binocular research that he was gracious in sharing with me. Dave Branchett, Phil Budine, and Gonzalo Vargas shared not only their binocular drawings, but also their enthusiasm, their drawing techniques, and some of their favorite objects. Likewise, one of my college students, Amber Hamilton, is a gifted artist and allowed me to use a couple of her sketches.

Even though they prepared what one would call technical sketches and diagrams, Ted Cox, David Frantz, and Rob Little contributed their incredible skills creating specific graphics for this book. I'd ask, "Can you illustrate this concept?" and it would appear—better than I had imagined! Wayne Green, one of my astro friends from junior high days, did an excellent job of illustrating binocular mounts with the help of a star trail, his wife, Christine, and a little computer assistance. Other photo contributions came from Dr. Wolfgang Wimmer of Carl Zeiss and Craig Weatherwax of Oceanside Photo and Telescope.

You will note my references to a couple of observing programs. My sincere thanks to Mike Benson, who chairs a number of the Astronomical League's Observing Clubs for his encouragement and support, as well as Elizabeth Waagen of the American Association of Variable Star Observers for graphics and program advice. I was also fortunate to have available the Washington Double Star Catalog, which is maintained at the United States Naval Observatory.

Some are talented with pen or pencil, others with photographic emulsion. The astronomical photos you see by Jerry Armstrong, Conrad Jung, Carter Roberts, and Vic Winter are simply some of the very best. I told someone that Vic simply "drips" with photographic talent, and I think you will agree.

I am also fortunate that I had a team of talented people who read the manuscript and provided solid input, on both the astronomy and the readability of the text. My most sincere thanks to Roger Curry, Dr. Rusty Harvin (another astro friend from elementary school days), Mike Ramirez, Mike Benson, and Pete Reynolds for their manuscript review and comments. My sister Susan LaForty worked hard trying to keep me grammatically correct. Any errors are my own and not due to anyone else.

David Levy is an amazing guy. I enjoy the opportunities to observe with David, who, like me, is enthusiastically rediscovering the sky with each look. I am humbled that he would write the Foreword for *Binocular Stargazing*.

Mark Allison and the entire Stackpole Books publishing team have again been a real pleasure to work with. They are supportive and committed to putting out a publication that readers will enjoy and find useful. I would especially like to thank Assistant Editor Chris Chappell, who has worked with me on this project.

Finally, most sincere thanks to my wife and best friend of more than thirty-two years. Debbie not only encourages my writing, but also is there to lift me up when I get bogged down. Plus, she still insists on taking those binoculars out when we go observing.

Keep looking up!

Mike D. Reynolds, Ph.D.
Jacksonville, Florida

Foreword

When you read about the latest discovery with the Hubble Space Telescope, you might think that the only things worth looking at are with the biggest, best, and most expensive equipment. I've got news for you: that is absolutely not true. For someone just starting a lifetime of discovery in the night sky, a pair of binoculars can be the best possible instrument.

The first optical system I ever used was my parents' 7×50 binoculars, a beautiful pair that they bought in the 1950s for their boat, the *Genie Pearl*, which was named after my two grandmothers. The stars looked like fireflies fluttering about. It took me a short time to learn that if I steadied the binoculars, the image would be much better.

Almost half a century later, though I still enjoy the *Genie Pearl*, my binoculars are no longer on a boat but strapped to a special binocular observing chair. And they give some of the best views of the night sky I have ever seen, no longer from the water but toward the ocean of space. The Andromeda Galaxy appears in all its glory, whereas through a telescope you could only see a part of it. The Pleiades looks better through binoculars than through any telescope. Even the Double Cluster in Perseus, a beautiful swarm of stars, reveals itself beautifully through binoculars.

What do I think is the most spectacular thing to look at with binoculars? It's not really a thing, but a journey that my wife, Wendee, taught me. On a nice, dark night, I love to cruise up and down the Milky Way. Like a beautiful country drive, the experience can be breathtaking. Beginning at the North American Nebula near the star Deneb, I head south through Cygnus, the Swan, stopping at interesting groupings of stars on the way. Then I get to a place where the Milky Way appears to split into two, with a main stream of stars that continues south and a shorter stream that looks like an off-ramp. I may use the exit, or I may continue southward to the magnificent clouds of stars that mark the center of the galaxy and Scorpius and Sagittarius. Each time I do this, it's a new voyage of discovery.

The sky is truly a wonder to behold, and this book will show you how little it costs to open its doors. A simple pair of binoculars is by far the least expensive way to get started on your journey to the stars. The book you are holding is written by someone with more binocular experience than anyone else I know. You couldn't be in better hands.

David H. Levy

1

Looking Up

What telescope should I buy? This question is often heard from people just getting interested in astronomy or looking for a present for a child. We are surrounded by department store telescopes promising Hubble-like views of the universe, and wonderful advertisements in astronomical magazines describe the superiority of one instrument over another. Yet the answer for the novice just getting started in astronomy probably should be to buy a pair of binoculars.

Why start with binoculars, especially if one's goal is buying a telescope? A number of reasons exist for such a purchase:

- A pair of binoculars of reasonable quality can be bought for well under $100; a telescope of reasonable quality can cost twice that much, or much more.
- Binoculars are easier to learn to use than a telescope.
- Objects are easier to find with a standard pair of binoculars than a telescope, allowing the novice to begin to learn the night sky and navigate around from object to object.
- If you decide that astronomy is not for you, you can always use the binoculars for other things.
- Two eyes are simply better than one.

Many amateur astronomers keep a pair of binoculars with them at all times when out observing. Binoculars can be useful for first examining a part of the sky before a particular object is located. And when that occasional fireball appears, a pair of binoculars is useful for examining the smoke trail, or train, often left behind— and, if you are quick enough, the fireball itself.

The low power or magnification and wide field of view—how much of the sky you can see when looking through an optical instrument—make binoculars ideal for the novice and even the experienced amateur astronomer. There is a misconception that high power in a telescope is the most important thing. And, in fact, many department store telescopes (and occasionally binoculars) are marketed by claims of amazing views at ridiculously high magnifications. The rule of thumb for telescopes is generally 50 to 60 power maximum per inch of primary objective. So for a 2-inch telescope, the maximum power under ideal conditions should be 100 to 120, not 575 as cited on the department store telescope's box. The bottom line here is that higher magnification does not translate into a better telescope.

Another important consideration is that binoculars are generally more portable and easier to use (just take them out of their case) than a telescope, which needs to be set up. Giant binoculars do require mounts for ease of use, but they are more like telescopes. Because a pair of binoculars is lightweight and takes up little space, it can easily be taken on trips, which might allow you, depending on where your travels take you, to see objects you normally cannot see from your home.

TWO EYES ARE BETTER THAN ONE

One of the major reasons for using binoculars is simply the human design: They feel more natural because we are created with two eyes. The word *binocular* is derived from the Latin for "two eyes." It's much more comfortable for the observer, especially the novice, to use both eyes. And there are real advantages to doing so.

Your eyes are the most used astronomical instrument. Human vision involves three steps: a geometrical optical step, in which light from an object is focused on the retina; a detection step, in which rod and cone cells in the retina detect incident light; and a data-processing step, in which electrical signals produced by stimulated detector cells are interpreted. This is accomplished through networks in both the retina and the brain, and also involves memory functions.

The eye focuses incoming light through the cornea (the principle refracting surface of the eye) and onto the retina. The lens, located behind the cornea, serves mainly to adjust the focal length

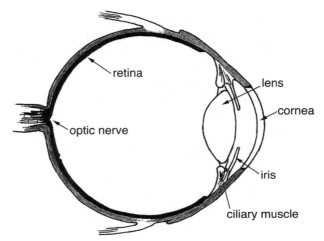

Parts of a human eye. ILLUSTRATION BY ROBERT LITTLE

of the combined cornea-lens system. The lens provides accommodation for distance; that is, it changes shape to bring images into focus on the retina for objects of varying distances from the eye. The shape of the cornea is normally fixed, although it can be altered through surgery or modified with a contact lens. The shape of the eye's lens—thus the focus—is controlled by the eye's ciliary muscle.

The human eye is most sensitive to light with a wavelength that is green in color (about 540 nanometers), but responds to wavelengths from violet (about 400 nanometers) to red (about 700). In the retina are detector cells, responsible for light detection. There are two basic types of detector cells: rods, which are rod-shaped, and cones, which are cone-shaped. Rod cells are found only in one type and provide only gray-scale vision, or black-and-white shades of gray. Rod cells are very light sensitive and function best in low-level or dim lighting, like most astronomical observing. Rod cells are relatively slow in sending signals to the brain.

Unlike the single-variety rod cells, cone cells come in three types: red, green, and blue. Red cone cells detect red light best but also detect, in order of sensitivity, orange, yellow, green, and blue. Green cone cells detect green light best, but also detect other wavelengths of light with varying sensitivity. And blue cone cells detect blue light best, but also detect some violet and green light. Even though they are sensitive to color, cone cells are not very sensitive to light, so they do not work well in low-level or dim lighting. Cone cells are faster than rod cells in sending signals to the brain. So cone cells provide us with day vision, and rod cells provide night vision.

Because of the distinct functions of the cone and rod cells, our eyes function in both very bright and very dim light. Human eyes make two adjustments to accommodate variations in light intensity: the pupils change size, and the retinas switch the type of detector cells being used. The pupils can vary from about 2 millimeters in bright lighting to about 6 or 7 millimeters in dim lighting. This is a change of roughly ten times the surface area, allowing much more light to enter the eyes under low light levels.

The retinas change the light into electric impulses that are carried through the optic nerves to the occipital cortex, the vision center of the brain, where the images are interpreted. The ability to see depth, or a three-dimensional view, occurs as a result of the brain's processing of the images from both eyes. The brain combines the two images into one, looking for similarities and differences.

The brain better processes images you are looking at with both eyes. Using both eyes to observe allows more light; some estimates are up to a 40 percent apparent increase in light. This translates to the ability to observe fainter objects. Resolution, the ability to see (or resolve) objects or distinguish between two closely aligned objects, is increased, as is image contrast, the ability to see and differentiate between fine details that are side by side. The ability to detect color is also enhanced in many observers. And using two eyes allows the brain to perceive almost stereoscopic images. For many observers, this stereoscopic effect is most apparent when viewing the moon, especially with larger binoculars.

All of this is affected by seeing, used here as an astronomical term that refers to the overall quality of binocular or telescope observations—the extent to which a star remains steady and free from twinkling or an image of a planet stays sharp. This is due to local atmospheric conditions that have little to do with how clear or cloudy it is—haze, for instance.

HOW BINOCULARS WORK

Although binoculars and telescopes both magnify objects, they differ in the way they work. Telescopes invert objects, turning them

1. Normal image

2. Reversed image

3. Inverted image

4. Inverted and reversed image

ILLUSTRATIONS BY DAVID FRANTZ; PHOTOGRAPHS BY MIKE REYNOLDS

upside down and sometimes reversing them left to right. Binoculars use a series of prisms to keep an image horizontally and vertically accurate. Adding the optics required to do so decreases the incoming light, which is important, especially when viewing fainter objects. But inverted images simply would not be acceptable for binoculars, especially for everyday uses such as sports, hunting, observing nature, or birding.

Binoculars are more like monoculars, which also employ prism systems. Monoculars are preferred by those interested in nature studies or birding who don't want the weight or bulk of a pair of binoculars. For astronomical purposes, however, a telescope is better than a monocular, which produces less light because of its prisms, and lacks the ability to increase or decrease magnification, an important aspect of telescope viewing.

A HISTORY OF BINOCULARS

To look at the history of binoculars, one needs to examine the history of the telescope. The telescope's invention occurred such that an exact time or location cannot be determined. Historically, it is usually associated with the Dutch eyeglass maker Jan Lippershey of Middelburg, Zeeland, in 1608. There are a number of written references to the telescope prior to Lippershey, however, yet no concrete evidence.

Historical records show that Lippershey might have built one or more pairs of binoculars (beginning in late 1608) in the process of applying for a telescope patent for a device designed for "seeing faraway things as though nearby." The instrument consisted of convex and concave lenses in a tube. If these binoculars were produced, it is believed they would have been around 3 to 4 power, with objectives of about 1.5 inches or less in diameter, making them something like 3×38s in today's binocular terminology. Quartz most likely would have been used for the objectives, since optical glass in that era was of very poor quality. And when one looks at the telescopes of the era—in particular, the Galilean optik stick—monocular viewing would have been awkward and with a very small field of view. The patent for Lippershey's binoculars was denied. Lippershey was later hired as a telescope maker for the state of Zeeland.

Galileo Galilei learned of the telescope and, building his own, apparently was one of the first to use it astronomically, in 1609. His crude instrument, by today's standards, allowed Galileo to discover the craters and mares on the moon, sunspots, the four largest moons of Jupiter (called the Galilean satellites), and the rings of Saturn, which he thought were akin to human ears.

Galileo's, and most likely Lippershey's, telescope used a combination of a double convex lens (outwardly curved on both sides) for the objective, the front lens focusing the incoming light, and a double concave lens (inwardly curving on both sides) for the eyepiece, the back lens focusing the light from the objective. This Galilean form of telescope presented the viewer with a very small field of view, even though the image was upright.

Just a couple years later, in 1611, German astronomer Johannes Kepler published an overview of a telescope that used double convex lenses for both the objective and eyepiece. The field of view was much larger than that of the Galilean telescope, but the image was inverted. This was not a problem for astronomical uses, however. Additional lenses could be added to reinvert the image, and this was accomplished by Anton Maria Schyrle. The difficulty here was that the additional lenses, not of the optical quality we are accustomed to today, diminished the incoming light, which impeded astronomical use.

Several people are known or believed to have ventured from the production of telescopes into binoculars or to have gone directly into binocular making. There are some indications that Galileo himself might have constructed a binocular helmet for use on ships at sea. Others include Ottavio Pinani in 1613; Antonius de Rheita in 1645, who first published a claim of inventing binoculars; and Capuchin monk Cherubin d'Orleans in circa 1670, said to be responsible for an ornamented cardboard binocular in the Firenze (Florence) Museum of Science in Italy.

Pietro Patroni of Milan, Italy, appears to have been a prolific telescope and binocular maker around 1700, and his instruments can still be obtained at auctions. Many of the optical and mechanical designs of these early instruments remain unknown, although it is believed that most used Galilean optics to produce an upright image.

One of the problems with all early telescopes and binoculars was the use of a single lens each for the objective and the eyepiece. There were a number of problems, the most serious of which was chromatic aberration. This occurred because the objective lens acted as a prism, breaking up the incoming light into its component colors. This might cause one to see, for example, a false-colored tint around the moon's circumference.

To get around the problem of chromatic aberration, Sir Isaac Newton used a concave mirror as the objective and a simple flat mirror to reflect the light out the side of the telescope to the eyepiece. This design, known as the Newtonian reflecting telescope, is still very popular today.

Another Englishman, Chester Moor Hall, devised a system that used two lenses of different types of glass for the objective to correct for the false color. Known as an achromatic refractor, this design also is still in use today. Other, even more advanced refractors use three objective lenses; these are known as apochromats and are very expensive compared with achromatic systems. The introduction of achromatic lens systems made possible better-quality binocular systems as well.

A number of early and interesting binocular designs have been recorded. Binocular theater opera glasses apparently were first produced in 1823 by Johann Friedrich Voigtlander in Vienna. These were simple Galilean-type refractors joined by metal bridges. Each eye tube and focus was independent from the other. The glasses were elegant, being finished in gilt and ivory. Significant binocular telescope innovations were accomplished by J. P. Lemiere in 1825. A twin Newtonian reflector binocular system was proposed by M. Vallack and described by the famous English astronomer Sir John Herschel in 1861. Herschel also notes a "stereo-telescope" built by A. S. Herschel in 1855, and a second such instrument was described in 1859 by M. Helmholz. The famous opticians and telescope makers Henry Fitz and Alvan Clark each produced one pair of binoculars for the U.S. Naval Observatory during the Civil War. Both of these historic pairs appear to have been lost. The firm of Alvan Clark & Sons later made several binocular telescopes.

Today's modern prism binoculars began with Ignatio Porro's 1854 Italian patent for a prism erecting system, using what is now known as Porro prisms. Porro continued his work, including the production of monoculars using his Porro prism design. Other early makers of Porro prism optics for instruments, either monoculars or occasionally binoculars, included A. A. Boulanger in 1859, Emil Busch in 1865, and Nachet in 1875. Unfortunately, poor-quality glass used in the optics coupled with poor design and production techniques resulted in failure in all of these cases.

Many of the early Galilean opera glasses were spectacular in decoration but poor in optics and mechanical construction. They found their way into use during the Crimean War of 1853–56 but were described as useless toys. Many of them were improved, however, using some sort of achromatic objectives and multiple lens eyepieces to correct the distracting chromatic aberrations. Yet there were still problems with limited field of vision—often referred to as "tunnel vision"—meaning that these optical instruments remained of poor quality. In spite of this, these relatively lightweight optics were popular.

Ernst Abbe, a German optical designer who was associated with the University of Jena and had close ties to optical manufacturer Carl Zeiss, displayed a prism telescope designed according to Porro's principles at the 1873 Vienna Trade Fair. Abbe's innovation

A pair of Zeiss binoculars from 1894. IMAGE COURTESY OF CARL ZEISS AND DR. WOLF-GANG WIMMER

for this instrument was centered on cementing the prisms. High-quality modern binoculars were first sold in 1894, a combination of Abbe's optical design and the manufacturing techniques of Carl Zeiss and chemist Otto Schott, a renowned glassmaker with whom Abbe helped found a glassmaking company. These antique binoculars produced extremely sharp views and are still considered by many to be one of the most attractive binoculars ever made.

A number of major innovations occurred throughout the years. In 1919, wide-field eyepieces were incorporated, which made binocular observing easier. The year 1933 saw the use of light metals instead of the heavier brass and zinc for binocular housings. Optical antireflective coatings based on the research of Alexander Smakula, which increased the light transmission of binoculars, were invented in 1935. And air-separated objectives were incorporated, which led to a reduction in size and improved image quality, especially with larger-aperture binoculars. Roof prism binoculars were first introduced in the 1960s, making more compact binoculars than those with the Porro prism system.

2

Purchasing and Caring for Binoculars

One of the first things you will notice when buying a pair of binoculars is a set of numbers, such as 6×35 or 7×50. These numbers represent two important characteristics of the pair of binoculars you are about to pick up and use. The first number is the power, or magnification, of the binoculars. In the two examples above, the numbers represent 6 power and 7 power. The second number is the diameter of each front objective in millimeters. In these examples, the numbers represent 35 millimeters and 50 millimeters. Using metric conversions, with 25.4 millimeter being approximately equal to 1 inch, the 35-millimeter binoculars are about $1\frac{3}{8}$ inches and the 50-millimeter binoculars are $1\frac{31}{32}$ inches, or very close to 2 inches, in diameter. Both of these cases fall way below the maximum magnification per inch of 50 to 60 power, so there are no problems from using too much magnification for the diameter of the objectives. This is usually the case for all binoculars.

As you will discover, binoculars come in a variety of sizes, configurations, and optical qualities. A number of important features and specifications identify binoculars as excellent, acceptable, inferior, or unacceptable.

Although their small size and weight make them ideal for their intended use, opera glasses generally are poor for astronomy. As a first pair, giant binoculars with 80-millimeter or larger objectives are probably not a good choice, but they would be an excellent second pair. For astronomical purposes, the size and characteristics of 7×50s or 8×50s make them ideal. The quality is not the best, but they are inexpensive—they can be bought at a

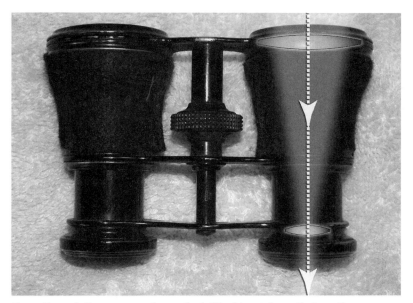

Even though these opera glasses look like binoculars, their optics are much simpler—the lenses are not achromatic and the optical path does not include prisms. They can be quite expensive, but they are not good for stargazing. ILLUSTRATION BY DAVID FRANTZ

department store for less than $50—and they will give you a good view of the night skies. And they are significantly better than anything Galileo Galilei used in 1609.

The larger the objectives, the more light-gathering power the binoculars have, which is generally good for astronomical viewing. This also means the binoculars are heavier, however, thus making observing more difficult or requiring support (not a bad characteristic, just one you should be aware of up front). Without support, every wiggle and vibration is more pronounced in higher-magnification binoculars. I like to have support for anything above 10 power or 70 millimeters in diameter. So bigger is not necessarily better; it all depends on what type of observing you would like to do.

Larger objectives also allow you to detect fainter objects. Astronomers use a system called the magnitude scale to indicate brightness. The magnitude scale was first introduced by the

DETECTING FAINT OBJECTS

Objective Diameter		Faintest Magnitude
In inches	In millimeters	
The Human Eye		About 6.5
2	51	10.3
3	76	11.2
4	102	11.8
6	152	12.7
8	203	13.3
10	254	13.8
12.5	318	14.3

natural philosopher Hipparchus in the second century B.C. Hipparchus ranked stars into six magnitude groups according to brightness. The brightest stars were first magnitude, the second brightest group second magnitude, and so on until the dimmest stars Hipparchus could see were sixth-magnitude stars. Modern measurements show that the difference between first- and sixth-magnitude stars represents a brightness ratio of 100. Thus a first-magnitude star is about 100 times brighter than a sixth-magnitude star. And each magnitude is $100^{1/5}$, or about 2.512, times brighter than the next. As a comparison, the sun on this scale is magnitude −26.7 and the full moon magnitude −12.5. This is also referred to as the apparent magnitude or brightness—what the objects look like from earth but not in true (absolute) comparison to each other. The human eye can see around magnitude 6 or 6.5 under dark skies, with good or corrected vision. By using a pair of 7×50s, you can see down to magnitude 10.3. And a pair of 100-millimeter binoculars will take you down to magnitude 11.8.

Another consideration is how the binoculars feel in your hands. Comfort while holding a pair of even relatively lightweight binoculars, especially for an extended period of time, is important. Many manufacturers have taken this into account for birders, hunters, and nature observers. Feel can include the overall shape—how well they fit in your hands—as well as the exterior coating. A number of manufacturers use a rubberized material or an armored body on some of their binoculars.

There are many more characteristics you should be aware of, especially if you are going to use the binoculars for astronomical purposes or are considering a pair of mini-giant or giant binoculars.

BINOCULAR SPECIFICATIONS
Classification
Binoculars are classified by the size of the objective lens (that second number after the ×). Even though manufacturers use different classifications, generally you will find the following classes or models of binoculars:

Mini or pocket binoculars. These have objectives not larger than about 25 millimeters and a roof prism, foldable/collapsible.

Compact binoculars. These have objectives not larger than about 25 millimeters and a Porro prism.

Standard binoculars. Objectives are greater than 25 and less than 60 millimeters, with a Porro or roof prism.

Mini-giant. Some manufacturers classify models that are 60 to 70 millimeters as mini-giants, typically with a Porro prism. These usually require some sort of tripod or specialized support.

50mm binoculars

80mm giant binoculars

100mm giant binoculars

PHOTO AND ILLUSTRATION BY DAVID
FRANTZ AND MIKE REYNOLDS

Giant. Binoculars with objectives of 60 millimeters or larger, or greater than 70 millimeters, depending on the manufacturer. These require some sort of tripod or specialized support. The largest commercial traditional binoculars on the market today are 150 millimeters (though some really dandy larger instruments are also available). Binoculars over 10 power sometimes are classified as giant, regardless of the diameter of the objective. Most manufacturers and dealers base this classification on the diameter of the objectives, however.

Military. You may occasionally run across binoculars labeled, "built to military specifications." If these are U.S. military specs, the binoculars are most likely okay. Some foreign militaries, however, view binoculars as disposable hardware.

Prism Systems

Quality binoculars use a series of prisms to both shorten the length (and reduce the weight) of the instrument and assure an upright and correct left-to-right image. You will find true binoculars with two prism types: Porro and roof. At first glance, roof prism (also referred to as compact roof prism) binoculars would seem superior to Porro prism binoculars because of their slender style and size. Porro prism binoculars, however, generally produce superior images that are sharper, brighter, and show better contrast. This is because of the way the prism systems are produced and function.

Roof prism systems are not as efficient as Porro prism systems at reflecting incoming light. Within a roof prism system, light reflects off at least one aluminized side, resulting in a loss of light, as reflective materials are not 100 percent effective. Thus objects appear slightly fainter. This is not important for daytime viewing, or even for bright celestial objects like the moon. But loss of light affects the viewing of very faint objects, especially those just at the edge of being seen, as well as objects that require the sharpest of images and best contrast. Some manufacturers use various technologies to minimize this loss of light; these enhanced optical systems are usually outstanding.

Roof prism binoculars are more expensive to manufacture because their more complex light path requires greater precision during the manufacturing process. Some roof prism binoculars are phase coated (or "P-coated") to correct for the loss of light inherent

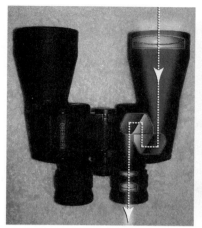 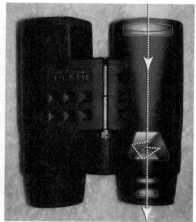

Porro (left) *and roof prism* (right) *binoculars, showing the light paths through the objectives, prisms, and eyepieces. The roof prisms allow for a more compact design.* ILLUSTRATIONS BY DAVID FRANTZ

in these systems. Phase-coated binoculars cost more but are worth the money for astronomical purposes if you decide to go with roof prism binoculars.

Roof prism and Porro prism binoculars have some obvious physical differences. Roof prism binoculars resemble two telescopes attached side by side; the light path appears to be straight through the optical system (though in reality, it is not). Porro prism binoculars exhibit that familiar binocular "zigzag" look because of the light path through the prisms. Reverse Porro prism binoculars are designed such that the objectives are closer together than the eyepieces.

Porro prisms can be produced from two types of glass. Barium crown glass, or BaK-4, prisms are far superior to borosilicate glass, or BK-7, prisms because they transmit nearly all the light through the system. This means brighter images through comparable pairs of binoculars, and for nighttime use, that's important. Most binoculars that use BaK-4 Porro prism systems state so on the binoculars, usually right below or above the power and diameter notation. If in doubt, there's an easy test: Hold the binoculars away from your body at about arm's length, looking at the eyepiece end. If you see perfectly circular light, they have BaK-4 prisms. With BK-7 prisms, the edges appear somewhat gray, almost diamondlike, because of the falloff of light.

Some instruments marketed as binoculars are not true binoculars because they do not contain prism systems. This is especially true of many opera glasses on the market today, as well as some of the "binoculars" sold for the sportsman or nature observer. These are in reality refracting telescopes with optics that correct the inverted images. Additional lenses tend to add internal reflections, especially noticeable if these lenses are not coated, or poorly coated at best.

Another important consideration is sturdy construction. Otherwise, if the binoculars are dropped or in some cases jarred, the prisms can easily be knocked out of alignment. If this occurs, the binoculars are basically unusable. It's difficult to observe with misaligned binoculars, even if the misalignment is so small that you cannot visually detect the problem. Your eye muscles will detect the misalignment and try to adjust accordingly, which can produce headaches. With most binoculars, losing their alignment requires you to return them to the factory. Some pairs of binoculars now include a function for the owner to realign the system. This is challenging but can be done.

Coatings

Lens coatings are another important factor to consider. An uncoated optic loses about 4 to 10 percent of the light striking it. Through the application of a thin coating of magnesium fluoride (MgF_2) on both sides of a lens, this is reduced to about a 1.5 percent loss of light. These lens coatings increase light transmission and reduce light scattering.

Hold a pair of binoculars at an angle to a light source and look at the front objectives from the side. You may notice a slight purplish tint, indicative of magnesium fluoride, as you move the binoculars. A whitish tint means the lens is uncoated, a pinkish tint means the coating is too thin, and a greenish tint too thick.

Manufacturers make all sorts of coating claims. Just because a pair of binoculars states it is coated doesn't mean the coating is of the highest quality. "Coated" usually refers only to the external lens surfaces, not the internal optical surfaces, and a loss of transmission and light scattering will still occur. If the binoculars advertise "fully coated" optics, this means every optical surface has been coated at least one time. Fully coated optics can reduce the loss of transmission and light scattering to less than 0.5 percent. By far the best for observing the night sky are "fully multicoated" optics, but they are also the most expensive. Coatings are best applied in several microscopically thin layers of magnesium fluoride rather than one thicker layer. You might also come across binoculars with "multicoated" optics, which means that there are multiple coatings only on some of the optics. Do not confuse these with binoculars that feature fully multicoated optics, which are superior.

Fully multicoated optics will also appear as a greenish tint when examined at a narrow angle to a light source, as described above. So how do you differentiate between fully multicoated optics and optics that have been too heavily coated? Look to see if the manufacturer states that the optics are coated, fully coated, or fully multicoated. The price is another indication; if it seems too good to be true, then the optics probably aren't fully multicoated.

Some binocular models sport ruby or red coatings or UV coatings (or UVC). Ruby coatings are said to reduce glare, which is not a problem for astronomical purposes. UV coatings filter out ultraviolet rays to reduce glare and protect your eyes from harmful rays, also not an issue when viewing the night sky. Occasionally UV coatings signify that the objective lens components—the two lenses—were coated on each side before they were cemented together.

Field of View
The field of view refers to how much of an area a pair of binoculars allows you to see. It is measured in degrees of arc or as a field width in feet at 1,000 yards from the observer.

For example, you might see listed as a specification, "Field of view is 350 feet at 1,000 yards." This tells you that through the lenses, you can see something 350 feet long that is 1,000 yards away. But for astronomical purposes, the field of view in degrees is most useful. The arc from horizon to horizon is 180 degrees, and your eyes can take in about 50 to 60 degrees of the sky. The moon is about 0.5 degree, or 30 minutes of arc.

Manufacturers might list the field of view in degrees, but is it the true field of view? I find it best to simply use the binoculars to look at the sky to determine the field of view. The bigger the field of view, the better. With a field of view less than 4 degrees, it becomes a little difficult to orient yourself. And if the field of view is too small, usually less than 2 degrees, you would need a finder scope as is used with a telescope. Some binoculars are marketed as wide-field-of-view instruments. These are usually a little more expensive, but this can be a nice feature for astronomical stargazing.

Today, binoculars' field of view can range from 1.75 degrees for Fujinon 40×150s to 9.50 degrees for Miyauchi 7×50s.

Exit Pupil

The beam of light leaving the eyepiece of the binoculars is called the exit pupil. The diameter of this beam ideally should match the diameter of the observer's pupil. This varies from about 2.5 millimeters in bright light to 7 millimeters in dim light. Astronomical observation is usually done under dim lighting conditions (except for the full moon and certain types of solar observing), and thus a larger exit pupil is more desirable.

You can hold a pair of binoculars at about arm's length and see the actual exit pupil. To determine the exit pupil's diameter in millimeters, divide the diameter of the front objective by the magnification. For a pair of 7×50s, you divide 50 by 7, which means that the exit pupil is a little more than 7 millimeters in diameter—excellent for observing the night sky! But a pair of 10×50s has an exit pupil of only 5 millimeters. A pair of 10×50s might still be excellent for someone who is older (usually your pupils' diameters get smaller as you get older) or if there is a lot of stray light from a bright moon or streetlights where you're trying to observe.

Some manufacturers use a different spec called twilight factor. Instead of stating the exit pupil diameter, they multiply the magnification times the objective diameter, then take the square root of this product. A pair of 7×50s has a twilight factor of 18.7, and 10×50s a twilight factor of 22.4. The claim is that the higher the number, the better you will see dim images. Other manufacturers may use terms such as relative brightness (RB) or relative light efficiency (RLE). For astronomical observation, do not worry about these characteristics; exit pupil diameter is the factor you want to consider along with other factors that determine what works best for you as an observer.

Focus

Focus is a major mechanical feature of a pair of binoculars. It can be accomplished through either individual eyepiece focus systems or a central focusing system. (These are referred to as "center-focusing." Avoid "fixed focus" binoculars, which don't allow you to change the focus.) Less expensive binoculars, especially some of the mini-giants and giants, usually employ individual eyepiece focus systems. There is nothing wrong with this as long as the focus system is smooth. And in fact, many people prefer an individual focus (IF) system.

Center focusing (CF) can be accomplished through several means, most commonly involving a knurled focusing disk, a rockerlike bar or plate, or a lever, with the eyepieces joined by what is called the eyepiece bridge. The lever system is not as good for astronomy; it is basically designed for people who need to continually change the focus as they track moving objects such as birds or race cars. For observing the night sky, once you have set your focus at infinity, it should be set for the evening, possibly with occasional fine adjustments.

On most center-focusing binoculars, one eyepiece can be adjusted to correct for different seeing in each eye. This system is referred to as diopter control. Choose an object and focus on it with the fixed eye first, usually the left eye, closing or covering your other eye while doing this. Once the left eye is focused, close that eye and adjust the focus for the right eye with the diopter control only, looking at the same object. Many diopter control focus systems have numbers so you can remember the setting.

There is a way to properly focus individual focus binoculars as well. On each eyepiece is a series of diopter numbers. Focusing one eye at a time, start at the highest positive number and move in gradually toward the negative numbers. Testing conducted during World War II showed that the focus becomes sharper when focusing this way; it has to do with how the eyes compensate during the focusing process.

The major complaint about the individual focus system is that when people share a pair of binoculars, each eye must be focused for each user. But center focus systems have two steps too, as you focus the entire system for one eye, and then tweak the focus for the other with the diopter control.

Hollywood versus reality: On the left is the binocular view we often see in television and movies, which is incorrect. The right-hand view gives a better picture of what the binocular stargazer will see. ILLUSTRATIONS BY ROBERT LITTLE

Interpupillary Distance

The distance from eye center to eye center of each person varies, usually between 2¼ and 2¾ inches. To accommodate this variation in interpupillary distance, binoculars are hinged so that the two halves will open and close slightly. This motion should be smooth and not bind at any time.

If you try to observe with the binocular set at the wrong interpupillary distance, you will not be looking through both eyepieces or seeing two distinct circles. This is a common error for nighttime observers, especially novices. Such people often wonder what everyone else is raving about after looking through the binoculars. Television programs often show binocular views as two merged circles; I chuckle every time I see this image. If your interpupillary distance is incorrect, that is what you will see.

Some people find it easy enough just to readjust the interpupillary distance with each use; others prefer to mark their interpupillary distance with a dot of paint or India ink. Many pairs of binoculars have an interpupillary distance scale that allows you to remember your personal setting.

Tripod Mount

For binoculars that are heavy, have a higher magnification (greater than 10 power without an image stabilization feature), or will be used for an extended period of time, a mount for a tripod is an important feature. The typical built-in tapped mount is located in the hinge of the binoculars and is a ¼-inch 20-threaded (¼-20) mount, a standard for photographic tripods. It is usually protected by a plastic cover or cap, which is unscrewed to access the mount. Some binoculars—especially giant binoculars—feature a bar with an adjustable mounting system. Such a sliding adjustable mount allows you to better balance the binoculars on a tripod or stand. Even if a pair of binoculars doesn't have a built-in mount, you can purchase an accessory that attaches to the binoculars and allows them to be tripod mounted (see chapter 3).

Image Stabilization
A number of manufacturers have begun producing binoculars that feature image stabilization, which dampens out most vibrations. Canon led the way with the Canon IS Series, a roof prism binocular system. Others with image stabilization systems include Fujinon and Zeiss. The mechanism for image stabilization varies among manufacturers. Image stabilization binoculars are expensive compared with other similar-size binoculars, even those of high quality. Many of those who own image-stabilized binoculars simply swear by them, whereas other users report a feeling akin to motion sickness.

Top right: *Magnification and field of view*

Top left: *Adjustable diopter focus with numbered indicators*

Bottom left: *Tripod tapped mount*

PHOTOS AND ILLUSTRATIONS BY DAVID FRANTZ AND MIKE REYNOLDS

Waterproofing

Many outdoor people, such as birders, hunters, nature lovers, and boaters, want binoculars that are waterproof, though they are not usually necessary for stargazing. These are not designed to be immersed in water, even though some manufacturers show ads of binoculars being held underwater. A waterproof pair of binoculars, however, will not be damaged by a little moisture or even fairly adverse conditions. Waterproof binoculars are usually rubber-coated optical systems sealed with rubber O-rings to prevent moisture intrusion. They are occasionally filled with dry nitrogen gas to prevent internal dewing.

Zoom

Zoom binoculars have a lever or knob that you can move to double or triple the magnification. You'll see these denoted something like 7–14×40 or 8–24×50. They may seem enticing, but they are not good for astronomical purposes. To produce the zoom feature requires additional lenses, degrading the brightness and occasionally the overall quality of the image. These binoculars are heavier as a result of the zoom's mechanical requirements. And the money spent on this additional feature could be spent elsewhere.

Other Features

Straps are important for hand-held binoculars and are excellent insurance against dropping them. You can replace the thin straps supplied with most binoculars with a thicker strap that is easier on your neck. Lens caps are important to keep the external optics clean when not in use. The binoculars case can vary from a simple pouch to an aluminum foam-lined case.

A number of binoculars offer rubber eyeguards or extendable eyecaps. The rubber eyeguards make easing into the binoculars while observing at night more comfortable, and they can also help block stray light. Some who wear glasses while using their binoculars note that the eyeguards help.

You might see other options as you research what's available. You will probably find extendable dew/glare shades useful for nighttime viewing. Another useful feature would be a set of filters, either attachable or built-in.

Some features, quite frankly, are useless for stargazing. One is a compass. If you need to determine which direction is north, use the stars. Another feature is a rangefinder, which tells you how far away something is. It might have a little difficulty with the Andromeda Galaxy!

SUMMARY OF BINOCULAR SPECIFICATIONS

Specification	Refers to	Notes and Preferences
7×50 (example)	The first number, 7, is the magnification or power. The second number, 50, is front lens diameter, in millimeters.	• The first and foremost identification factor in choosing a pair of binoculars • The larger the lens, the more light-gathering power (and the more expensive and heavier)
Magnification or Power	The number of times the object appears enlarged	• 7 power is good for general stargazing • 10 power and above is good for solar system observation as well as splitting some double stars
Diameter	Size of the objective, in millimeters	• Smaller objectives, around 50 millimeters, are good for general stargazing • Larger objectives, giant binoculars, are better planetary binoculars but also can be excellent for deep-sky observing
Prism system	Porro prisms versus roof prisms	Porro prism systems are generally much better for astronomical purposes; lower cost
Coatings on the optical surfaces	The type of coatings: none, coated, fully coated, multicoated, and fully multicoated	Fully multicoated optics are, by far, best for astronomical purposes
Field of View	How much you can see, measured in degrees of arc or as a field width in feet at 1000 yards	Wider is better

SUMMARY OF BINOCULAR SPECIFICATIONS (continued)

Specification	Refers to	Notes and Preferences
Exit pupil	The diameter of the beam of light exiting the binoculars' eyepiece; determine by dividing the objective diameter by the power	Usually the larger the exit pupil, the better. The best range, depending on the age of the observer, type of observing, and observing site, is about 5 to 7 millimeters
Eye Relief	Distance from the eye to the binoculars to see the entire field of view	Best is around 15 to 20 millimeters • Too short: uncomfortably close • Too far: hard to hold steady and center
Focus	Individual eyepiece focus system or center focus	Individual eyepiece focus systems are less expensive; most important factor is smoothness of focus
Interpupillary Distance	The distance between the observer's eyes	An adjustment which allows for the binoculars' eyepieces to be separated
Tripod Mount	A ¼-inch 20 threaded (¼-20) mount to allow a tripod to be attached	Usually found on giant binoculars; a must if the binoculars need to be tripod mounted
Image Stabilization	Motion/vibration is damped out	A nice, but expensive, feature; some people get motion sickness by using image stabilized binoculars
Waterproof	Usually water resistant (not truly waterproof)	For adverse conditions, not usually necessary for stargazing
Nitrogen-filled	Sealed binoculars filled with dry nitrogen gas	Nice option that keeps interior moisture-free; expensive
Zoom	Binoculars that allow for a variation in power instead of a fixed magnification	A single lever or knob can double or triple the magnification; not a good choice for skywatching

BUYING BINOCULARS

When you decide to buy that first (or second) pair of binoculars, you will have a number of options to choose from. Whether you don't want to spend much or want to buy the best, the market has a lot to offer. For relatively little money you can buy a decent pair of binoculars. Not top-of-the-line binoculars with BaK-4 Porro prisms, fully multicoated optics, and outstanding eye relief, but an adequate pair that will give you a good start at binocular stargazing.

Although the largest selection is available through mail order, many department stores offer a variety of binoculars at reasonable prices. I visited two well-known, large national chain stores to see what types of binoculars they carried, as well as the prices. One store carried about sixteen different kinds of binoculars, the other eighteen. The brands included Bausch & Lomb, Bushnell, Nikon, and Tasco. You might be surprised to find Nikon in this group; I was too. They were not Nikon's best binoculars. Selling for $140, they were 10×50s featuring BaK-4 prisms but were only multicoated, not fully multicoated. For $25, one could purchase a pair of Simmons Red Line 10×50 binoculars. These binoculars have BK-7 Porro prisms and coated optics, not the best features, but still a pair of binoculars that will allow you to begin to learn about and enjoy the night skies.

I was a bit surprised to find high-magnification binoculars that featured only 50-millimeter objectives, such as 12×50s and 16×50s. These are not good general binoculars to own. The higher magnifications mean a reduced field of view, making it harder to find objects, and a tripod is necessary to keep them steady if you want to enjoy observing through them. I don't recommend buying anything over 10 power unless you are going to mount it on a tripod.

You will probably find some of the better brands and models at local stores. Sporting goods shops and camera stores often carry binoculars made by camera manufacturers such as Canon, Minolta, and Nikon. Meade Instruments, well known for its telescopes, also makes several models of binoculars. Meade's Montana series of binoculars, 7×42s and 10×42s with fully multicoated optics, are simply outstanding. They are comfortable to hold, and even with roof prisms (silver phase coated), they deliver terrific views comparable to those through binoculars that cost five to seven times more. I was very pleased with images of the moon

As this wall display makes clear, there are a great variety of binoculars on the market today. PHOTO COURTESY OF CRAIG WEATHERWAX, OCEANSIDE PHOTO AND TELESCOPE

through the Meade Montana 10×42s. Locally, you might also run across binoculars made by Celestron International, another well-known telescope manufacturer.

If you can wait, you will probably do better via mail order, in terms of price and selection. There are a number of excellent options available; see appendix B for a starting list. Before ordering binoculars through mail order, make sure they can be returned if you find the images unacceptable or the construction poor.

Some might also look at eBay or other online auctions. A recent search on eBay turned up nearly a thousand pairs of binoculars being offered. These included everything from Fujinons and Zeiss on down to quite unacceptable "binoculars." As always with online auctions, buyer beware! Make certain you read the fine print, including shipping and handling. One pair of binoculars I looked at was being offered at an opening price of 99 cents. This may sound like a real bargain, but the shipping and handling on these inferior binoculars was $70, making this "deal" not so special.

Another option, especially for used binoculars and equipment, is a website called AstroMart. Here you will find listings of all types of used and new astronomy equipment for sale. The binocular section usually has around thirty to fifty pairs of binoculars for sale, as well as accessories. Note the recommended guidelines by AstroMart to assure a smooth transaction.

A number of manufacturers import binoculars specifically for amateur astronomers. These are made in Europe or the Far East. In particular, Chinese-made binoculars are making quite a splash in the United States among amateurs because of their low prices. You can get some really good binoculars for the money. Several companies have distinguished themselves in this regard. Most of these binoculars are available only through mail order.

The Chinese-made Oberwerk binoculars are quickly becoming a mainstay among amateur astronomers, birders, and nature watchers. Oberwerk produces a number of models, from 8×42s with phase-coated roof prisms to a number of 100-millimeter binoculars. You might not be able to afford a pair of the giant Fujinons, but you should look at the giant Oberwerks. For $400, you can get a pair of 22×100s featuring BaK-4 Porro prisms with fully multicoated optics. Or if you would like a solid pair of binoculars, look at the Astronomy Binoculars, an investment at $1,500. One model even has eyepieces set at a 45-degree angle for ease of viewing. This is an investment that you will enjoy for many years.

The most popular pair of Oberwerk binoculars appears to be the 15×70s, and for good reason. They feel great when you hold them, although at 15 power, they really need to be mounted on a tripod or binocular mount. Like the 11×70s, they have BaK-4 Porro prisms with fully multicoated optics—for $150. When I tested a pair of the 15×70s against what many would consider the tip-top binoculars (at nearly ten times the price), I was surprised and pleased at how well the Oberwerks compared. If you are looking for a solid pair of "standard" binoculars, the Oberwerk 8×56s, also featuring BaK-4 Porro prisms with fully multicoated optics, cost $100 and produce images that are sharp and show good contrast and resolution.

Apogee Instruments is another Chinese company that offers a variety of binoculars at good prices. One of the neatest options from Apogee is a series of six models called AstroVues. These binoculars, ranging in size from 7×50s up to 20×100s, have built-in dark sky filters that are inserted or removed with the flick of your thumb.

The European-made binoculars, like Zeiss, are nearly always of high quality—and usually expensive. These are usually considered family heirlooms and are often passed from generation to generation. Other European manufacturers that you might not recognize include Leitz and Steiner. An Italian manufacturer, Astromeccanica, produces fine binocular refracting telescopes in 120- and 150-millimeter sizes.

A few manufacturers, such as Miyauchi and Takahashi, make binoculars that are apochromats rather than achromats. These have a three-lens objective system (versus the achromatic two-lens system) that corrects for all chromatic (color) aberrations found in such achromats. You will pay a premium for these, as with apochromatic refracting telescopes. And you will have to decide if the additional expense is worth the nearly perfect colors in the images.

There are dozens of manufacturers and distributors that retail thousands of binocular models. It will amaze you what's on the market today. Just know that all binoculars, regardless of where they are made, are not made alike. There have been some complaints about poor optical quality or uneven manufacturing standards with some of the foreign-made binoculars. Ask around, check websites and see what people are saying, or join one of the Internet equipment or binocular threads or chat rooms. There are plenty of people who are more than willing to share their observations—not only astronomical—with you. A little patience and research will produce that pair of binoculars that's ideal for your purposes.

So what do you need in a pair of binoculars for astronomical purposes? It depends on how you intend to use the binoculars. I recommend at least a pair of 7×50 BaK-4 Porro prism fully multicoated binoculars with center focus and an eye relief between 15 and 20 millimeters. If you want to reduce expenses because of a tight budget, you could buy a pair that features BK-7 Porro prisms or is fully coated instead of fully multicoated. But if you compare them side by side, you will be able to tell the difference.

If you want to purchase a pair of mini-giant or giant binoculars, a number of options and models are available. You can spend a reasonable amount (less than $400) or a significant amount (more than $10,000). Giant binoculars must be mounted for viewing ease, however; holding a pair of 100-millimeter binoculars and enjoying the view simply is not an option. The good news is that there are some outstanding options for mounting a pair of giant binoculars, and some of these mounts make viewing an absolute pleasure.

The most important criterion is a piece of equipment that you will use—and will want to continue to use. If you find that a pair of binoculars is difficult to focus, too heavy for you to hold, or has other problems, it will probably wind up on a closet shelf a few years after you bought it, having seen little use. So get a pair of binoculars you will want to use—and will enjoy using—to explore the night sky.

CLEANING YOUR BINOCULARS

Once you've invested in a pair of binoculars, especially if you purchased a high-quality pair, you'll want to keep them clean. This is not only to keep them looking nice, but also to protect your investment and continually provide you with the best view possible. There can be up to 50 percent reduction in light through dirty optics. But no matter how hard you may try to avoid it, if you use your binoculars, they will get at least some dust on the outer optical surfaces. And mascara and lash cosmetics can get on the exterior eyepiece lenses.

Try never to touch the exterior optics, and never simply rub a lens or use a paper tissue on it. A tiny amount of dust or a grain of sand can cause a scratch if rubbed. The oils in your skin can cause degradation of many optical coatings.

Begin cleaning by blowing off the lenses with a squeeze bulb or compressed canned air (make certain the compressed air will not leave a substance on the lens); do not blow on the lenses with your mouth. A photographic camel's hair brush can also be used to gently brush across the optical surfaces.

Once you are absolutely certain the lens is dust- and grit-free, you can use a lens-cleaning solution along with a grit-free photographic tissue or wipe if grime, fingerprints, eyelash oil, mascara, or other stains remain. The solution should be applied to the tissue, never directly to the lens. Wipe with gentle motions across the lens. A dry tissue might be needed to clean off any residual cleaning fluid.

There are other options for careful optical cleaning. Some people prefer to use pure isopropyl alcohol, acetone, or ethyl alcohol instead of lens-cleaning solution. Make sure whatever you use is absolutely pure, or it can leave stains on your optics. Your local pharmacy is a good source for these items. Instead of photographic tissues or wipes, some people prefer cotton balls or swabs; others like photographic or optical cloth wipes. The only cautions are to make certain that any cleaning materials are clean—it's best to purchase optical cleaning materials—and not to rub the optics.

3

Accessories and Specialized Products

Once you have chosen and purchased a pair of binoculars, there are some accessories and tools that will make exploring the night sky much easier and more pleasurable. In some cases, they are almost a must.

If you are new to astronomy, the first things you should buy are a planisphere and a red-filtered flashlight. A planisphere, or star wheel, is a flat star chart with an overlay that represents the horizon. By turning the star wheel, you can set the planisphere for a given day and time for the latitude. Most planispheres show the constellations and identify the bright stars.

A red-filtered flashlight will allow you to see the planisphere at night so that you can identify constellations, stars, and other astronomical objects. A white light would cause your eyes to shift from night to day vision, thus reducing your ability to see faint objects in the night sky. A red light allows you to see the planisphere without affecting your vision.

You also eventually should purchase a set of star charts or a star atlas. I recommend using star charts that do not go to very faint magnitudes; 50-millimeter binoculars allow you to see down to about the tenth magnitude.

Once you are set with the basics, there are some other accessories that can make your viewing more pleasurable. The optical manufacturers and others have designed and built some spectacular products that parallel binocular observing or take it a step further. Some of these are a matter of common sense; others demonstrate incredible ingenuity.

DEW PREVENTION

In some places, dew can present a problem for sky watchers because it coats the exterior of the binoculars or telescope. Dew forms when an object or surface cools to a temperature that is lower than the dew point of the surrounding air. The dew point is both temperature- and humidity-dependent and is usually given in degrees. Once the temperature of the object drops below this point, dew forms. Dew is made of liquid water that has condensed, or formed droplets, from the water vapor in the air. If an entire layer of air cools next to the ground, then fog forms.

To prevent dew from fogging your objectives, you can use dew shields or dew caps, extensions of the objectives' lens cells. These are easy to make and can be as simple as rolled cardboard fitted onto the ends of the objectives. A few manufacturers, such as Orion, offer extendable dew and glare shields on some binocular models. But sometimes the dew point is such that no matter how long the dew shields are, dew still forms on the objectives. And the eyepieces on your binoculars are also exposed and susceptible to dew.

One of the many binocular accessories available is a blow dryer that plugs into a car's cigarette lighter. You can use it to warm the binoculars, especially the objectives and eyepieces, to a temperature above the dew point. Some manufacturers sell a heater band that goes around the front of the objectives to warm them.

FINDERS OR SIGHTS

Sometimes it is useful to have some sort of finder scope on your binoculars, especially if they are higher power. A finder or sight may be handy for the novice regardless of the binoculars' magnification or field of view. And even more seasoned astronomers use them, as a simple sight makes finding an object much easier. Many of us use such devices on our telescopes, especially the big Dobsonian reflectors; you'll see them marketed under names like Telrad. A number of binocular finder options are available from different optical manufacturers, or you can pick up a finder in the sporting-goods section of a local store. These "point sights," made for airguns and other similar guns, require you to make a dovetail mount. A sight or finder for your binoculars should be low power. You will often see finders or point sights marketed as 1×. This means there is no magnification whatsoever, which can be useful for binoculars.

TRIPODS AND MOUNTS
For binocular stargazing to be enjoyable, it should not be taxing. And one of the primary ways to avoid a tiring session is to mount your binoculars. This is a must for heavy binoculars, generally those over 10 power and those classified as mini-giants or giants. Let's face it, it's next to impossible to hold a pair of 20×100s up to your eyes for any length of time.

Camera tripods are the simplest systems for mounting binoculars. Your binoculars must have a built-in ¼-20 thread mount, however, or you will need a binocular-to-tripod mount accessory. Even with the ¼-20 thread mount built into the binoculars, you will need another device to go between the binoculars and tripod: a binocular tripod mount. One end of this L-shaped device attaches to the binoculars and the other to the tripod. If you are using this type of mounting with heavier binoculars, you will probably need a heavy-duty tripod, as a steady and sturdy tripod is essential. When using an instrument that magnifies an object, such as binoculars or a telescope, vibrations are also magnified.

Although a binocular tripod mount can be an inexpensive way to mount your binoculars, it is not the easiest system to use. You'll find viewing at certain angles difficult, especially above 45 degrees. Looking horizontally or near the horizon presents no difficulties, but you have to become a contortionist to look overhead. And it's difficult for more than one observer to share the mounted binoculars, especially if their heights are significantly different.

For this reason, several manufacturers have developed some outstanding mounts made specifically for binocular stargazing. One of the older systems is known as a double parallelogram. I've used one of these for about fifteen years, and you might still see them offered on the used market. They work better than the camera-binocular tripod mount combination, as they are much more stable and easier to use. On the downside, they require a heavier mount than does a camera tripod, and some parts of the sky are not easily accessible. They are bulky and heavy, take up a lot of space, and can be difficult to set up. And you must use them from a standing position or with a special observing chair.

A pair of 70mm mini-giant binoculars with an Orion EZ finder, all attached to a Blaho binocular mount

Other mounts and systems have been designed to allow comfortable viewing of the entire sky. Many of these mounts and their manufacturers have come and gone, yet you might find a source that still stocks mounts from a company that is no longer in business or locate one of these mounts for sale used on AstroMart or eBay.

The Blaho Company manufactures a single parallelogram binocular mount, available in two models: the Stedi-Vu and the Stedi-Vu Junior. Options include a 1x finder and tripod head. The finder does not magnify but simply gives you an easy aim at the sky. This is probably not necessary in low-power binoculars, but it might be useful for a higher-power pair. The Stedi-Vu mounts, constructed from black aluminum bar stock, use a counterweight system with Teflon bearings for smooth motions. A well-made tripod for the Stedi-Vu is a must, since azimuth rotation—the movement along the horizon—depends on the tripod.

Burgess Optical offers red oak binocular mounts in three sizes, all single parallelogram design. Burgess also makes two tripods, one made of furniture-grade red oak and the second described as a "magnesium" tripod.

From Orion Telescopes and Binoculars, the Paragon-Plus binocular mount is a single parallelogram design with an extendable counterweight. It can be purchased either with or without a tripod and can be used for binoculars up to 80 millimeters.

T & T Binocular Mounts offers a number of products. The company's most popular mount is the ARTiMount, designed to allow you to sweep through 110 degrees horizontally without any motion except for your head. This mount can also be used for seated observations. The Kids Mount, a single parallelogram design made of solid oak with an adjustable counterweight, supports lightweight binoculars up to 80 millimeters on a camera tripod that you supply. T & T also manufactures a mount specifically for wheelchairs, called the Wheelchair Mount.

T & T also has a series of tripods made from crutches, both wooden and aluminum models. If you are handy and would like to try to make a crutch tripod of your own, this is an interesting concept to consider.

Here, Christine Green demonstrates the ease with which binocular mounts move. ILLUSTRATION BY WAYNE GREEN; PHOTOS BY WAYNE GREEN AND MIKE REYNOLDS

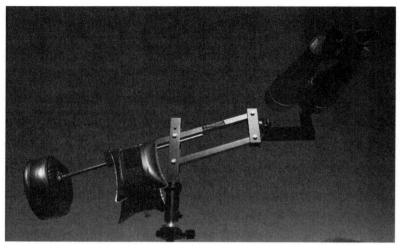

This is the author's setup for taking giant binoculars on long trips; it includes the Universal Astronomics UniMount and Apogee's 100mm AstroVue binoculars. Note that the supplied weights have been replaced with water weights, which makes for a lighter load when traveling.

Another major source of high-quality binocular mounts and accessories is Universal Astronomics. Universal Astronomics has a number of models, including the ever-popular UniMount, T-Mount, and UniStar. These mounts are made from 1-inch-square aluminum tubing with Teflon spacers between the beams, resulting in extremely smooth motion. The UniMount and T-Mount are single parallelogram design, whereas the UniStar is an alt-azimuth type mount, meaning that there are two axes of rotation: altitude (or vertical) and azimuth (or horizontal). The beauty of the Uni-Mount is that it is comfortable to use whether you are seated or standing. The mount is easy to set up, use, and disassemble. An adjustable counterweight system makes movement very smooth, and the binoculars stay in place when you stop moving them. Even though the UniStar can be used with a pair of binoculars, it is better suited for a telescope. I recommend purchasing the UniStar only if you plan to use it for a telescope in addition to binoculars. Universal Astronomics also has a variety of tripods for use with its mounts, including custom L-shaped tripod mounts, as well as extra counterweights and other accessories. All of these products can be purchased directly from Universal Astronomics or through an authorized dealer.

Virgo Astronomics used to manufacture binocular mounts but is no longer in business. You may still find some of these products available, such as the Nova/mount and Sky/mount.

Regardless of the type of binocular mount you use, you must have a solid tripod. The weight of the binoculars and mount combined is often fairly heavy and can be too much for many of the lightweight camera tripods. And when you move the binoculars around, a less-than-solid tripod will be shaky, and so will the objects you are trying to observe.

Other options include chest-mounting your binoculars, with a strap placed around your neck and the mount resting on your chest. Don't forget to give your own body sufficient support as well. If you are using binoculars that do not require a mount for an extended period of time, make certain you will remain comfortable. I prefer to use a chaise lounge chair, which allows me to lie back at an angle from which I can comfortably observe the sky.

MIRRORS

A novel idea for ease of viewing is using a mirror for binocular observing. This idea is not a new one; in the late 1950s, when sky-watchers were using telescopes to look for satellites, many of their instruments were pointed down at a mirror, which reflected the image into the telescope's objective.

Ease of use is a major goal with many observers, and some feel that looking comfortably down into the eyepieces of the binoculars while seated is easier on the back and neck. But the mirror and mount system must be of high quality, or your image will suffer serious degradation. As in reflecting telescopes, the mirror should be "first surface," which means that the reflective coating, usually aluminum, should be on the exterior glass surface and not on the back. Mirrors used in bathrooms, automobiles, and for other common uses are "second surface." The problem with second-surface mirrors is that you get two reflections: one faint reflection off the uncoated glass, and a second, brighter reflection off the reflective surface. The mirror must be thick enough that it doesn't bend or warp the image. Warping can also occur if the mirror is not optically flat.

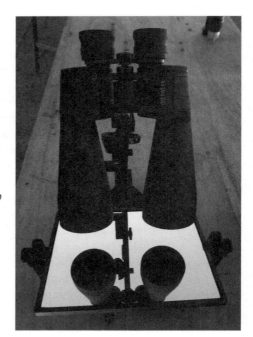

Another type of mount, the Sky Window, with a pair of Oberwerk 70mm binoculars. This is an excellent option for those who wish to be seated while observing.

The mount that holds the mirror should not put strain on the mirror but must support it, allow for the mirror to be adjusted (to look at the sky), and be able to hold a pair of binoculars. One such product is the Sky Window, manufactured by Trico Machine Products. The Sky Window consists of a first-surface, ¼-inch-thick mirror on a solid mount of aluminum. An overcoat protects the reflective coating. Mirror motions are smooth. The Sky Window can be used on a tripod or table. The observer is seated while looking down into the eyepieces of the binoculars. You will have to get used to the image, though; the orientation of what you see is different, as it's a mirrored view. You also will lose a little light; Trico says the mirror reflects 95 percent of incoming light, a loss of 5 percent. But for some observers, the ease of observing offsets these small drawbacks.

FILTERS

The use of filters on telescopes for a variety of purposes is common. Different filters are available for different types of viewing.

Color filters can really enhance the viewing of the moon and planets. These come in standard colors, and some are also numbered (such as Yellow #12). You will be surprised how much difference using a yellow filter makes while exploring our moon. And there are special filters for observing comets, too. Deep-sky observers (i.e., those observing objects beyond our solar system) consider many filters, such as those that filter out light pollution and nebular filters, part of their standard equipment.

Not all binoculars can easily take a set of filters. Some observers have made special adapters to allow for the fitting of filters between the exterior eyepiece lenses and the eyes. Whether this will work depends on the type of rubber eyeguard or eye shade on your binoculars, if any. One manufacturer, Apogee, uses built-in dark-sky filters in its AstroVue line of binoculars.

FILTERS AND THEIR USES

Filter	Use
Solar Filter	Reduces light to a safe level for viewing • Always place before the objective lenses; never between the eyepieces and the eyes *Note:* Viewing of the sun is dangerous and can cause permanent eye damage
Hydrogen-alpha or H-α	A special filtering system for solar observing • Narrow bandwidth allows viewing of solar prominances
Neutral Density or ND	Excellent for reducing the moon's intense light • Available in varying transmissions *Never use for solar observing*
Color Filters (often use Wratten Numbers such as Red 25A)	Helps improve definition of planetary features • Red, orange, yellow, green, blue, and violet are the most common colors
Oxygen III or OIII	Excellent for planetary and bright nebulae • May help with light pollution
UHC	Excellent for all types of bright and diffuse nebulae • May help with light pollution

BINOCULARS FOR SOLAR OBSERVATION

Specialized binoculars have been manufactured specifically for solar observations. Coronado offers the BinoMite, 10×25 roof prism binoculars used for observing the sun in white light. The filters are hard-metal coated to prevent degradation, such as pinholes and other defects. All harmful radiation is safely blocked. Even though the binoculars have a small diameter and lack Porro prisms, they are excellent for observing the sun—which is the only object you can observe with these binoculars. BinoMites are lightweight and perfect for solar eclipses, except during totality.

Coronado has taken solar observing one step further with the BinoMax Canon binoculars. A specially matched set of narrowband filters is integrated into Canon 12×36 IS (image-stabilized) binoculars to allow for observing the sun in hydrogen-alpha light. This narrow wavelength allows you to see prominences, solar flares, filaments, and other solar chromosphere details not visible in white light—with both eyes. BinoMax Canons provide simply stunning views of the sun. They are expensive, and solar observing is the only thing you can do with these binoculars, but for some, this is enough.

TELESCOPE BINOCULARS AND VIEWERS

Do you covet a really big pair of binoculars but can't afford the big 150-millimeter Fujinons? Well, here's some good news: Jim's Mobile Incorporated (JMI), which has been developing and manufacturing equipment for amateur astronomers for many years, has recently introduced the JMI RB Line of reverse reflector telescope binoculars. These telescope binoculars, available with 6-inch (150-millimeter) and 10-inch (250-millimeter) primary mirrors, are carefully manufactured and assembled to assure solid collimation, alignment so that the two parallel telescopes are looking at the exact same point in the sky.

JMI is not the first to attempt to build reflector telescope binoculars; a couple other manufacturers have made twin reflecting telescopes over the years. I once used a pair of 10-inch reflector telescope binoculars made by another manufacturer, but collimation between the two telescopes was difficult. And without proper collimation, I got a headache. But when I could get the two telescopes collimated to each other, the views were incredible.

JMI has worked hard to overcome these difficulties, and it seems to have succeeded. I had an opportunity to look though JMI RB Line reflector telescope binoculars recently and found them simply incredible.

Several manufacturers have developed telescope binocular viewers, which substitute a binocular view for the monocular view. Binocular viewers take the monocular light from the telescope and, using high-quality prisms, split the light into a binocular viewer; two eyes are used to view the object instead of just one. These instruments produce views that are out of this world! The moon appears three-dimensional, and many deep-sky objects take on a completely different appearance. Telescope binocular viewers are expensive; even the lower-priced models can cost as much as a good telescope. Binocular viewers differ from binocular telescopes in that only the telescope's eyepiece is being substituted for the viewer—the viewer does not require two telescopes mounted side-by-side.

I once asked a group of some twenty nonamateurs to tell me what they thought about viewing though a telescope binocular viewer on my Meade LX-200 8-inch telescope. Most of them thought the telescope binocular viewer made a major difference, but three had more difficulty seeing the image, even a first quarter moon. Once I showed them how to adjust the binocular viewer, as well as how to use it, two of the three had no additional problems. It is sometimes difficult for a person to see through a telescope's

Left: *The Denkmeier II BinoViewer on the author's Meade 8-inch LX-200 GPS Schmidt-Cassegrain telescope.* Right: *Denkmeier's Power x Switch, which adds more versatility to the BinoViewer.*

eyepiece, as those of us who set up telescopes for general public viewing have experienced.

A number of manufacturers offer telescope binocular viewers, including Burgess Optical, Celestron International, Denkmeier Optical, Lumicon, Tele Vue, and University Optics. Prices range from $200 to $1,000. It is clear that the higher-priced models produce better images; you can find a number of opinions on Internet threads about this subject.

The Denkmeier and Tele Vue models come with various options to enhance the viewing experience. These will add cost to the base price but are worth the additional investment. For example, the Denkmeier offers the Power × Switch, a wonderful accessory that allows you to change magnification simply by the pull of a lever, kind of like a pair of extra-giant zoom binoculars. The Power x Switch also allows Schmidt-Cassegrain telescope owners to reduce the magnification through a focal reducing lens, resulting in less power and a wider field of view.

4

Lunar and Solar Observing

The moon and the sun are great places to start observing celestial objects with your binoculars. If you've never seen a magnified view of our nearest celestial neighbor, the moon, you will be amazed. And yes, there are safe ways to view the sun, using proper filters or projection. *Warning: Never view the sun directly through any optical instrument, including binoculars. It can cause permanent eye damage.* This chapter describes some safe methods as well as what you can see, including the beauty of lunar and solar eclipses, some of nature's most awesome events.

THE MOON

In 1609, Galileo became the first person to observe the moon with a telescope. He must have been amazed at what he saw. Today the binoculars you use, even an inexpensive pair from a department store, will provide you a better view than what Galileo experienced. And you too will be amazed.

The moon is one of the most fascinating objects in the sky, and you can see a myriad of details even through a pair of 7×50s. It is interesting to follow the changes in phases with binoculars as the moon revolves around the earth. Simply following the play between lunar day and night can also be intriguing. As you observe the moon from night to night, you will be amazed how its face seemingly changes. What was clearly visible one night can be difficult to see the next.

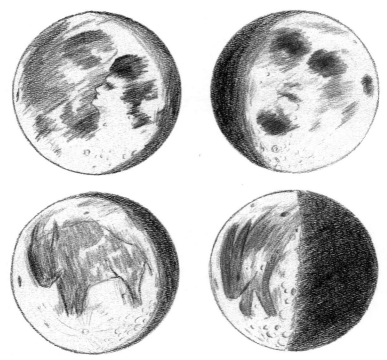

Objects we "see" in the moon, clockwise from top left: lady, man, rabbit, and buffalo ILLUSTRATIONS BY ROBERT LITTLE

Lunar Features

You might be overwhelmed at all the features you can see on the moon. Begin by learning to identify some of the more prominent ones, and you will then be able to locate the more difficult ones. Lists and even entire books have been written describing lunar features. A recent work by the highly respected Charles Wood, *The Modern Moon*, details the "Lunar 100," a list of lunar objects that starts with the easiest objects and becomes increasingly more difficult. Several of these objects are visible even with 7×50 binoculars. The list begins with L1, the moon itself. L2 is earthshine, and L3 differentiates between lunar mare and highland.

Another excellent list of lunar features visible through binoculars is the Astronomical League's Lunar Club, one of its Observing Clubs (see Appendix C). The list is divided into three sections:

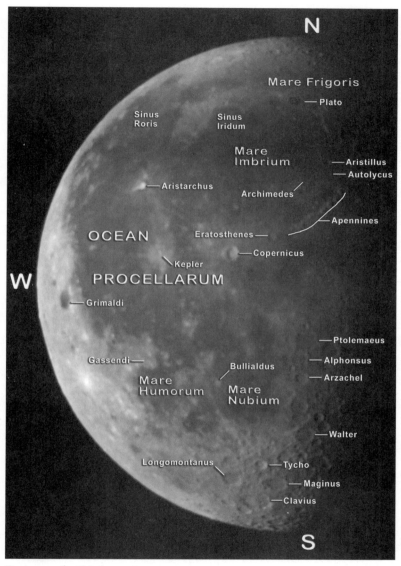

Features of a third-quarter moon ILLUSTRATION BY DAVID FRANTZ; PHOTO BY MIKE REYNOLDS

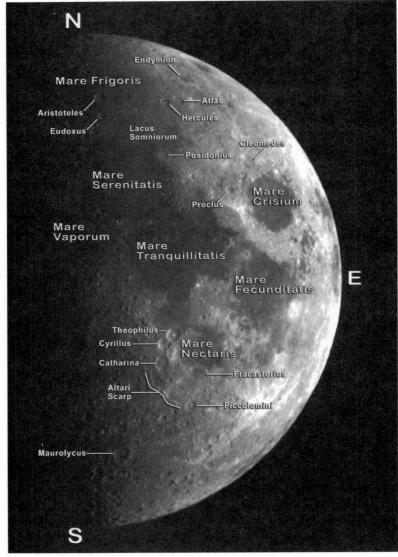

Features of a first-quarter moon ILLUSTRATION BY DAVID FRANTZ; PHOTO BY MIKE REYNOLDS

naked eye, binocular, and telescopic features. You will also find in appendix E a list of lunar binocular features by type and when each is visible during the lunar month.

The two photographic charts provided in this chapter will help you identify most of the basic features with binoculars. Because you are observing with low power, you will not see the detail that you would through a telescope. As you get more into lunar observing, it will become obvious that higher magnifications are necessary to resolve all those wonderful details. Giant 25×100 binoculars will open new views unseen with the 7×50s. If you get bitten by the lunar bug, you may want to graduate to a telescope, which will offer higher power and better looks at the moon's features.

One question often asked is whether you can see the flag left by the astronauts on the moon. Even with the most powerful telescopes, this is not possible. The following are the basic features you will be able to see on the moon's surface.

Terminator. One of the first things you will notice is the sharp division between day and night on the moon. The boundary between dark and light is called the terminator. Following the terminator night after night, starting right after the new moon, will unveil new lunar terrain. Features near the terminator are sharp because of the way the sunlight strikes the moon at this point; shadows are longer, and thus features are more pronounced.

Highlands. Highlands are bright mountainous features. They are generally heavily cratered and cover about two-thirds of the side of the moon facing Earth.

Maria, or mares. Galileo named these flat, dark lava plains maria (singular mare, the Latin word for "sea"). By examining the number of craters on the maria versus other lunar features, astronomers have determined that lava flows occurred after the moon had largely been fashioned.

Craters. Craters are circular indentions, mostly caused by meteoritic impact. They vary in size from microscopic to tens of miles across. It is rewarding not only to identify various craters, but also to see how small a crater you can resolve with your binoculars. This will depend on the phase of the moon, which dictates the overall lighting. It is most difficult to view craters and other features during a full moon, when the sun is basically at its highest

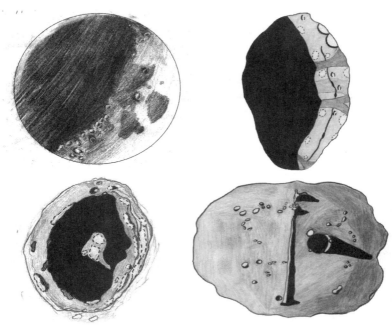

Some examples of lunar drawings, clockwise from top left: five-day-old moon, 36%; Aristarchus; Straight Wall; Theophilus ALL ILLUSTRATIONS BY PHIL BUDINE EXCEPT TOP LEFT BY AMBER HAMILTON

point in the sky as reflected on the moon, producing no shadows. Plus the full moon is bright, and the intense light can be hard to look at with optics. Filters can be used to decrease the light's intensity.

Ejecta and Rays. Crater ejecta is formed when the shock of impact uplifts rock under and near the crater, which then falls back, forming the crater walls. Imagine dropping a rock into a tray of fine flour. The flour will splash, or be ejected, around the area where the rock lands. The same effect occurred when some lunar craters were created. Rays are the brights streamers of rock that are thrown from the crater during impact. They can be seen as streaks that extend some distance from the craters and are best observed when the sun is directly overhead. The crater Tycho exhibits a wonderful system of rays.

Rilles. Rilles are cracks in the lunar surface. They make fascinating subjects but are difficult to observe with low-power optics.

The most famous and best example of a rille is the Straight Wall, a wonderful fault that needs to be caught at a specific time in the lunar cycle.

Lunar Eclipses

Binoculars can be an excellent means of enjoying a partial or total lunar eclipse. Lunar eclipses can occur only during full moon. There are three types of lunar eclipses: penumbral, partial, and total. Penumbral occurs when the moon passes into the earth's lighter penumbral shadow. Partial occurs when the moon passes into the earth's lighter penumbral shadow and the earth's umbral shadow, but not deeply enough to become completely immersed. Total occurs when the moon passes into the earth's umbral shadow deeply enough to become completely immersed. There are a number of events you can observe during a lunar eclipse.

Detection of the penumbral shadow. The penumbral shadow can be difficult to detect, especially during the early phase of a penumbral eclipse. Some observers have not noted the penumbra until just prior to the beginning of the partial lunar eclipse.

Timing the umbral shadow contacts. A good task is to time the moment when the earth's umbral shadow first seems to touch the moon, called first contact. Likewise, you can time the fourth contact, when the umbral shadow appears to last touch the moon. If you are observing a total lunar eclipse, also watch for the second contact, when the moon is completely immersed in the umbra, and the third contact, when the moon leaves the umbra. Use an accurate timepiece, and note the time when each contact occurs. You need to keep the binoculars very steady to make accurate timings; it's best to use a mount, even for 7×50s.

Crater timings. These timings can be made as the umbral shadow sweeps across the moon. The Association of Lunar and Planetary Observers publishes a list of specific craters for timing. Most observers time a central point; for large craters like Copernicus and Tycho, many time the umbra's arrival first at one rim and then at the opposite rim, later taking the average of the two. As with contact timing, you need to keep the binoculars very steady to make fairly accurate crater timings.

Color and shading variations. Variations in shading of the penumbra can often be noted; it is not uniform and usually becomes darker the closer it gets to the umbra. You also can watch

Partial lunar eclipse, October 27, 2004. Note the Earth's umbral shadow on the moon.

for color change from a full "bright white" to a dusky or yellowish brown in the penumbra, and differences in both color and shading in the umbra.

Lunar transient phenomena. During an eclipse, the moon undergoes a significant reduction in the amount of sunlight falling on its surface, leading to sudden changes in temperature. It is believed that these changes lead to stress, producing short-lived flashes of light, glows, and patches of haze.

Evaluating and estimating the color of totality. During a total eclipse, is the moon copper, red, grayish? A view through binoculars, even 7×50s, will help you make such an evaluation. Some observers keep a color wheel nearby to compare what they see with specific colors.

Estimate the brightness at midtotality. The midtotality is the middle of the eclipse. You can estimate the moon's brightness at midtotality using the Danjon Luminosity Scale, a simple scale developed by Andre-Louis Danjon (1890–1967). See the table on the next page for the specific categories.

Time stellar lunar occultations during totality. It's always fun to watch a star disappear or reappear from behind the moon, which is called a lunar occultation. Occultations occur because the moon is orbiting the earth. As it moves across the sky, it occasion-

Total lunar eclipse, October 27, 2004

ally moves in front of stars, causing the occultation. Lunar occultations can occur during a lunar eclipse. It is usually very difficult to observe when a star is occulted during full moon. But during a lunar eclipse the moon is darker, and so the star will be easier to see. And you might be in the right place at the right time and observe a graze, where a star passes tangent to the moon's limb or edge, blinking off and on as the moon's mountains and valleys pass in front of it. As with contact and crater timings, you need to use a mount for your binoculars. With lower power binoculars, the star will need to be fairly bright for you to even be able to see it.

THE DANJON LUMINOSITY SCALE

Category	Characteristics
1	A very dark eclipse. The moon is almost invisible, especially at midtotality.
2	A dark eclipse, gray or brown in color. Lunar details distinguishable only with difficulty.
3	Brick-red color. The umbral shadow usually has a bright or yellowish rim.
4	A very bright copper-red or orange eclipse. The umbral shadow usually has a bright, bluish rim.

THE SUN

The sun can provide observers with a fascinating object. The first and foremost concern is safety: Under no circumstances should you ever look at the sun with a pair of binoculars, except during the total phase of a solar eclipse. You must use proper solar filtering for both binocular objectives or project the image onto a white sheet of paper or cardboard. You can mount your binoculars on a tripod or binocular mount, point them at the sun, and then project the image. Never look through the binoculars to confirm that you are pointing at or near the sun. To find the sun, allow the shadow of the binoculars to become at a minimum. This means that your binoculars are more directly in line with the sun. You can use a small pinpoint finder mounted in parallel with the binoculars to make aiming at the sun easier and safer.

Once you have the sun in the binoculars and projecting through the eyepieces onto the white paper or cardboard, carefully place a cardboard screen at the front objectives to reduce the light around the projected solar images, making them easier to observe (see the photograph on the next page). In any case, do not allow your binoculars to stay focused on the sun for long periods of time. They can heat up, causing problems as extreme as lens breakage.

The white light you see is the sun's face, or photosphere. During active solar periods, you should be able to see black dots; these are sunspots. These areas appear black because they are cooler than the surrounding photosphere. There also are specialty binoculars that allow viewing the sun in hydrogen-alpha light.

Solar Eclipses

With the proper safety precautions, a pair of binoculars can be an excellent way to view a partial or total solar eclipse. Binoculars mounted for eyepiece projection work great for partial phases of an eclipse, allowing you to safely and easily see the eclipse. During totality, it is safe to view the sun directly through optical devices, including binoculars.

There are several types of solar eclipses, where part of the sun is covered or the moon is centered over the sun:

Partial eclipse. A partial eclipse occurs when the Earth-moon-sun alignment is such that only part of the sun is covered. A partial phase also occurs before and after a total or hybrid solar eclipse.

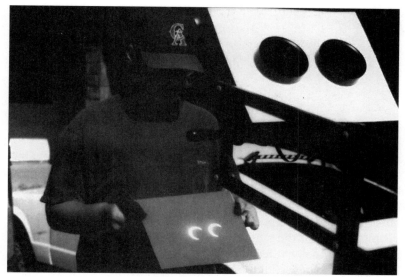

During the partial solar eclipse phase of the May 10, 1994, annular eclipse, Jeremy Reynolds holds a card on which a pair of 80mm binoculars is projecting two images of the eclipse. Note the card at the objective end of the binoculars, blocking sunlight from the projected image.

There are a number of events you can observe during a partial solar eclipse.

Contact timings. As with lunar eclipses, you can observe and time when the moon first makes contact with the sun, also called first contact. Second and third contacts denote the beginning and end of a central eclipse; fourth contact is the last instant you can see the moon "touching" the sun. Contacts can be timed by using an accurate timepiece and noting when each event occurs. Very steady or mounted binoculars are highly recommended.

Sunspot contact timings. As with lunar eclipse crater timings, you can time when the moon contacts any visible sunspots.

Baily's Beads and the diamond ring effect. Just prior to second contact and again before third contact, you will see some fascinating events. Baily's Beads look like brightly lit and irregularly spaced points of light seen along the edge of the sun just prior to and after totality. They are caused by the last few rays of the sun shining through lunar valleys along the moon's edge. The diamond ring

The brilliant diamond ring of the total solar eclipse on April 8, 2005   *An annular or "ring" eclipse of the sun* PHOTO BY VIC WINTER

effect is one brilliant point of light appearing along with the now-visible corona of the sun. Baily's Beads indicates that it's time to remove your solar filters; when you see the diamond ring effect, it's time to look away and replace the filters.

Annular, or ring, eclipse. An annular eclipse, also known as a ring eclipse, occurs when the Earth-moon distance is such that even though the eclipse is central, the moon, because of the distance from Earth, is not large enough to cover the sun. An annular eclipse can also occur as a result of the relative Earth-sun distance. Such as eclipse is still quite spectacular, and some aspects of totality, such as Baily's Beads, can be seen.

Total eclipse. One of the most spectacular astronomical events occurs when the moon completely covers the sun, known as totality. During totality, there are a number of things you can observe. Binoculars are simply terrific for all-around totality observations as the sun's visible face, the photosphere, is covered. You can look at the completely eclipsed sun with no protection. Use that precious time during totality to take it all in. Some astronomers conduct rigorous photographic sequences; others examine specific phenomena. The eclipse will be more memorable if you take the time to enjoy all aspects of totality. Many people, including me, believe that a total solar eclipse is the most spectacular of all astronomical events and feel grateful for a single opportunity to be in the shadow.

Corona. The most obvious aspect of totality is the sun's beautiful corona, the solar atmosphere beyond the chromosphere. The

The June 21, 2001, total solar eclipse. This image is a composite of ten photos taken at varying exposures to attempt to capture the full range of what the eye would see.

shape of the corona depends on solar activity. Look for long, filamentlike streamers near the sun's equator and short, fan-shaped appendages near the poles.

Chromosphere. Difficult to observe, the chromosphere is the reddish to scarlet layer of the solar atmosphere just beneath the corona and above the surface.

Prominences. Jets of gas that erupt from the solar photosphere are called prominences; they can take various shapes and be quite spectacular.

Sunrise-sunset effect. Brilliant red and orange horizon colors occur just before, during, and after totality. You might also want to use your binoculars to scan for the shadow as it approaches and then departs at the end of totality.

Stars, planets, and other phenomena. It will amaze you how dark the sky becomes during totality. Use your binoculars to sweep the sky, looking for bright stars and planets. Charts are available that show where various objects should be visible.

Some people use this opportune time to look for previously undiscovered comets near the sun. These comets otherwise would remain undetected in the sun's glare for weeks more before they moved far enough away from the sun to be discovered.

Hybrid, or total-annular, eclipse. In a hybrid, or total-annular, eclipse, the moon is centered on the sun, and part of the path of the central eclipse is annular—the moon is not quite large enough to cover the moon—and part of the path is total. Imagine a line that represents the central eclipse crossing Earth, and because part of Earth is fractionally farther away from the moon, it produces an annular eclipse.

5

The Solar System

Solar system objects are our nearest neighbors and are often some of the brightest objects in the sky, next to the sun and moon. Through binoculars, they can be some of the most rewarding or most frustrating to observe.

Galileo's first look at some of the objects in our solar system proved to be exciting and enlightening. He noted that Jupiter had four moons, which cast doubt on Ptolemy's theory of an Earth-centered solar system and supported the Copernican sun-centered solar system. He also noted that Saturn had "ears." Galileo's observations were made through a telescope that was of much poorer quality than a pair of modern 7×50 binoculars.

The issue with observing the planets with binoculars is mostly magnification. Even with 25-power, 100-millimeter binoculars, you will not get a spectacular view of Saturn's rings or Martian polar caps. For this you need larger aperture and higher magnification. But simply tracking these objects, such as the four bright moons that orbit Jupiter or some of the brighter asteroids, is quite doable in binoculars. For observing bright comets, and even some of the fainter ones, binoculars can provide a spectacular view, sometimes the best one. You can also use binoculars for meteor watching, often considered only a naked-eye type of observation.

ASTEROIDS

Asteroids (meaning "starlike") are small, rocky bodies that orbit our sun. They are also referred to as minor planets. Giuseppe Piazzi was the first to discover an asteroid, Ceres, in 1801 while searching for a planet in the large gap between Mars and Jupiter, where astronomers believed another planet orbited. There are thousands and thousands of asteroids, ranging in size from a few miles across to large ones like Ceres, which is about 637 miles "in diameter." Diameter holds a strange meaning for these objects, because the majority of them are irregular in shape, most likely the leftovers from the formation of our solar system.

Even with the best Earth-based telescopes, astronomers see asteroids only as starlike objects. Asteroids are found by detecting their movement against the background of fixed stars. This is one way that you too can find some of the brightest asteroids. About sixty asteroids become brighter than tenth magnitude during their orbit around the sun, so they should be visible in 50-millimeter binoculars, which allow you to see down to about magnitude 10.3. To look for asteroids, you need a good set of star charts and predictions for the location to plot on the chart.

Begin by locating the star field where the predictions show an asteroid to be. Make certain to confirm that it is the correct star field. Note stars that are near the location of the asteroid, and look back and forth between the star chart and the view through your binoculars to try to pick out a starlike object not on the chart. Because asteroids move against the background of stars, following the object over time will help you confirm that you have indeed identified an asteroid.

The *Minor Planet Bulletin* is the quarterly publication of the Minor Planet Section of the Association of Lunar and Planetary Observers. This publication provides updates on asteroid information and observations others have made.

COMETS

Few objects in the night sky interest and excite people more than comets. Comets are really just snowy dirtballs. They are made of carbon dioxide and water ices, ammonia, organics, charcoal, and sandlike substances.

A bright comet generally appears about once a decade, but many faint comets are discovered annually. Historically, comets were feared objects, often thought of as bad omens. The Bayeux Tapestry, depicting the battle of Hastings in 1066, shows Halley's Comet over the battle. The comet's appearance actually inspired the forces of William the Conqueror prior to their invasion; it was believed to have been an omen portending the death of Harold II.

Comet orbits are often narrow and sometimes described as cigar-shaped. The time it takes an object to go around the sun in its orbit is known as its period. Earth's period is 365.24 days, or 1 year. Short-period comets have periods of 200 years or less, and their orbits tend to lie relatively close to the plane of the solar system. The shortest-period comet currently known is Encke's Comet, with a period of 3.3 years. Halley's Comet has a period of 76 years. Long-period comets are those with periods greater than 200 years. Comet Kohoutek, visible in 1973–74, has a period of 80,000 years. Long-period comets have orbits that are random in shape, and they can come in from all directions.

Comets are believed to be from two general areas in our solar system: the Oort Cloud and the Kuiper Belt. The Oort Cloud is at the far extremes of the solar system, believed to be 20,000 to 100,000 astronomical units (AU) from the sun (an AU equals the distance from Earth to the sun, or 93 million miles). There may be trillions of icy objects in the Oort Cloud. These objects are pulled from the cloud by occasional passing stars, which change their orbits, sometimes enough to send a few of these bodies toward the sun. As a general rule, comets from the Oort Cloud have very long periods.

Halley's Comet, April 28, 1986 PHOTO BY VIC WINTER

Comet Ikeya-Zhang, March 31, 2004 PHOTO BY VIC WINTER

The Kuiper Belt, a closer region of icy bodies, lies just beyond the gas giant planets (Jupiter, Saturn, Uranus, and Neptune) and within the plane of the solar system. Some of the outer solar system objects, such as Chiron (which is cometlike in nature), Neptune's moon Triton, Pluto's moon Charon, and Pluto itself, might actually be or once have been Kuiper Belt objects. Thus a controversy has erupted over whether Pluto actually is a planet. The Kuiper Belt's icy bodies would explain short-period comets because they are closer to the sun. These bodies were probably formed from the solar nebula (the cloud from which our solar system formed), far enough from the gas giant planets to have survived.

The major parts of a comet are its head, or coma; a nucleus of material hidden within the coma; and the tail. The nucleus of a comet can be up to tens of miles across. The coma of a comet is a huge cloud of dust and gas. Spectral analysis shows that the gas of the coma is made of carbon, hydrogen, nitrogen, and oxygen atoms in ion and molecular form.

Comets can have two types of tails: gas or dust. Gas tails, called type I tails, appear to be fairly straight, with streaks and irregularities. They are made up of ionized atoms that have been released from the nucleus. Gas tails are most intriguing; they can disassociate from the comet nucleus, which then forms a new gas tail. Dust tails, or type II tails, appear to be rather smooth, with little or no visible features. The dust is very small—about 1 micron or a millionth of a meter in size—and reflects sunlight. Whereas a gas tail is blown relatively straight by the solar wind, a dust tail is usually arced because of the differential velocity between it and the coma and nucleus.

The type of tail or tails formed is particular to each comet. In 1996, Comet Hyakutake exhibited a spectacular type I gas tail, but no dust tail. Comet Hale-Bopp, in 1997, one of the most spectacular comets of the twentieth century, exhibited both a spectacular blue gas tail and a yellow dust tail. But research has shown that not all comets have tails; in fact, there is evidence that the majority of comets do not have visible tails. These comets apparently have exhausted their ices.

As a comet approaches the sun, the comet's nucleus, the concentration of the cometary mass, begins to heat up and "outgas," releasing gas and dust from the nucleus. This gas and dust first envelop the nucleus, then are swept behind the comet. This sweeping effect is due to solar wind, which consists of ions and atoms from the sun's corona that are moving rapidly outward through the solar system. Because of the solar wind, a comet's tail always points away from the sun. As these comets travel through the inner solar system, they leave a trail of material behind, somewhat like the proverbial trail of crumbs. This material is mostly dustlike in composition. Meteoritic dust is the source of several annual meteor showers, such as the Perseids meteor shower in August, caused by dust left behind by Comet Swift-Tuttle.

Observing comets can be a lot of fun. If the comet is bright enough, it is easy to find with binoculars. Comets that are at the limit of seeing with the naked eye or fainter are best found by using a star chart and a plot of the comet's location. Often comets, especially the fainter ones, do not have tails and appear as fuzzy objects. Brighter comets, on the other hand, may have spectacular tails. Binoculars are better than a telescope for observing brighter comets; the telescope's field of view is too limiting, and the magnification too high. I observed Hyakutake, Hale-Bopp, Halley, West, and other bright comets with binoculars and seldom used a telescope.

You can easily contribute data about a comet, such as its magnitude and size, to organizations such as the Association of Lunar and Planetary Observers. To estimate the magnitude, find stars near the comet that appear to be about the same brightness. Put these comparison stars slightly out of focus in your binoculars so that the "star blobs" look about the same size as the comet. Determine which star blob looks about the same brightness, find that star on your chart, and you have an estimate of the comet's magnitude. Knowing the field of view of your binoculars allows you to estimate the size of the coma and the length of the tail, if visible. Let's say the field of view of your binoculars is 5 degrees. The coma and tail appear to be about one-half of the field of view, so the length of the comet is 2.5 degrees.

Comet hunting is a challenging aspect of astronomy that you can attempt. Big binoculars are terrific for this; a number of comets have been discovered this way. Comet hunters search in the west just after sunset and the east just before sunrise. Many observe up to three hours after sunset and three hours before sunrise. The sky needs to be as dark as possible. The hunter sweeps parallel to the horizon (western in the evening sky, eastern in the morning), moving the binoculars along in a smooth path. Once the end of the sweep is reached, the observer moves the binoculars up one-half to three-quarters of a field of view, then sweeps in the opposite direction; some sweep all the way from the north to the south. The comet hunter has to be dedicated; it may take years before that first discovery is made, if at all.

Your heartbeat may increase when you see a fuzzy, cometlike object, but you must make certain that your discovery is not a

When comet hunting, the observer begins at the horizon and overlaps with each horizontal sweep. ILLUSTRATION BY DAVID FRANTZ

deep-sky object; it would be embarrassing to announce the discovery of an already known galaxy! Check Charles Messier's list, which he compiled to catalog all of the known objects that might resemble comets (see appendix G). So let's say you've found what you believe is a comet. You've confirmed that it's not an already known comet or a deep-sky object. It's not some sort of internal reflection from a nearby bright object. You've patiently noted that it is moving against the background of stars. What do you do now? You contact the Central Bureau for Astronomical Telegrams (CBAT). But be as certain as possible of your observation before doing so. CBAT notes that for every confirmed new comet discovery, about five more "discoveries" are reported that are not comets. If you are the first to discover a new comet, the reward might be a celestial object with your name. The International Astronomical Union has established specific guidelines for naming comets.

METEORS

Most meteor watching, either casual or during a specific meteor shower, is done without the aid of optical devices. Meteors can appear from all over the sky. Even during a shower when they appear to radiate from one specific point, these meteors will cover a larger area.

Occasionally observers see a meteor while looking through a telescope or pair of binoculars; these optics allow you to see much fainter meteors than with the naked eye. Such meteors are called telescopic meteors and look no different than naked-eye meteors. The advantages are that telescopic meteors appear brighter, you are able to see fainter meteors, and you may have the opportunity to view the meteor's train, a smoke trail left by some meteors, if one is produced.

Telescopic meteors are more difficult to catch, since a telescope or pair of binoculars significantly reduces your field of view. Nevertheless, some observers take on the task of making telescopic meteor observations. Usually a pair of binoculars or a short-focus, wide-field telescope is chosen for telescopic meteor work; low magnifications are preferred, and this gives binoculars the edge.

There also has been evidence of telescopic meteor showers, ones that are visible only through a telescope or binoculars.

The 2001 Leonids meteor storm as seen near Ayers Rock, Australia. This time-exposure photo shows hundreds of meteors coming from the storm's radiant. PHOTO BY VIC WINTER

PLANETS AND THEIR SATELLITES

Observing the planets with binoculars is a challenge, and to some, it may be disappointing. Use the maximum magnification possible: 20×, 25×, even 30× binoculars if you have them.

Mercury. Mercury is the most challenging to observe of the five naked-eye planets, because it never is very far from the sun. When visible, Mercury can be seen just after sunset or right before dawn. Binoculars are handy to search for Mercury; check predictions for visibility.

Venus. Venus is often called Earth's twin, and depending on where it is in its orbit, it is sometimes the planet that is closest to the earth. Like Mercury, Venus can be seen just after sunset or right before dawn. Venus is brilliant when visible and is often called the evening or morning star because of its brightness (only the sun and moon are brighter). Since Venus (as well as Mercury) is closer to the sun than is Earth, it appears to go through phases akin to the moon, which can be observed through binoculars. This was one of the pieces of evidence Galileo used to confirm a heliocentric solar system.

Here Jupiter and its Galilean satellites are seen through two different sets of binoculars. On the left are Jupiter and its four brightest moons as seen through 7× binoculars. On the right is the view through 20× binoculars.

Venus can also be seen transiting the sun through binoculars. You must always use caution in observing the sun, however (see chapter 4). The last transit of Venus in our lifetime will be June 5, 2012; these rare events occur in pairs (the last transit of Venus was June 8, 2004), and these pairs occur at roughly 122-year intervals.

Mars. When Mars, the Red Planet, is close to Earth or in opposition (when the Earth-Mars distance is at its minimum), there is usually a lot of interest in it. Binoculars, even at 30× magnification, will not provide you a good view even at the closest of oppositions, with the exception of a tantalizing glimpse of its reddish color.

Jupiter. Jupiter is the king of the planets, with its own solar system. At the higher magnifications a pair of binoculars can provide, you will be able to see some of the features on the cloud tops of this fascinating world. Bright zones and darker belts are visible, though details will not be. The four bright Galilean satellites—Callisto, Europa, Ganymede, and Io—will be visible as starlike objects to either side of Jupiter. All four might not be visible at once; charts showing the daily locations of these moons relative to Jupiter are available in a number of publications.

Saturn. Your first glace at the ringed planet, Saturn, through binoculars will most likely be somewhat disappointing. You will experience what Galileo experienced: the frustration of not being able to resolve the rings of Saturn, for they are just out of reach of a

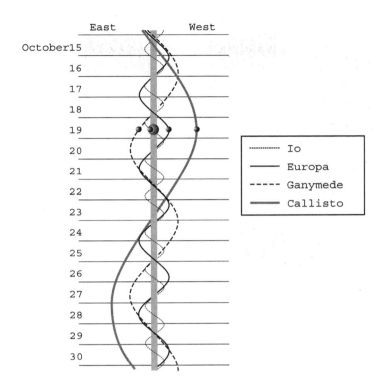

Jupiter Moons Chart

A number of magazines and books contain monthly charts that show Jupiter's four brightest moons relative to the planet as they rotate around it. ILLUSTRATION BY DAVID FRANTZ

30× pair of binoculars. You can, however, catch glimpses of Saturn's largest moon, Titan.

Uranus and Neptune. These two outer worlds in our solar system are good challenges for binocular observers. Uranus is barely visible to the naked eye. Both look starlike through binoculars and are a challenge even for earth-based telescopes. You will need a star chart and predictions to locate Uranus and Neptune, but each is worth the trouble in your exploration of the solar system.

The moon's passage through the skies and the position of a given planet will occasionally bring the two near each other, and the moon may occult the planet. Shown here are Jupiter (left) *and Saturn* (right). PHOTOS BY VIC WINTER

Pluto. A debate rages over whether Pluto is really a planet; it seems cometlike in nature. Because of its extreme distance and small size, Pluto is around a very faint thirteenth magnitude in brightness, thus not something you can spy with binoculars in the night sky.

ARTIFICIAL SATELLITES

Even though they are recent man-made additions in our solar system, observing artificial satellites through binoculars can be a lot of fun. These earth-orbiting objects are seen just after sunset and prior to sunrise, usually moving generally west to east. Unless you have a go-to telescope with a function that allows you to track artificial satellites, binoculars are the optics of choice because of the low magnification and field of view. It is easy to observe several satellites in an evening; there are a number of sources for predictions of the specific locations.

In the early days of the space program, many people were part of Moonwatch teams, as they were called, to track artificial satellites. The brilliant Echo 1 communications satellite was the first artificial satellite I saw through binoculars. Echo 1 was no more than a large, reflective balloon, but it made a wonderful target for a young boy's eager eyes, even with inexpensive binoculars.

6

Stars

As you look up in the night sky, you will casually note that the stars do not all appear to be the same. Even at first glance, it is apparent that these diamonds of the night are of different brightnesses, which astronomers measure as the stars' magnitudes. Stars are rated on what is called the apparent magnitude scale. A star can appear to be bright because it truly is bright in comparison with other stars, or it can be relatively dim while appearing to be bright because it is very close to us. If, however, we compare "apples to apples," theoretically placing all objects the same distance from the observer's point of view, then determining their brightness, we have what is called the absolute magnitude scale.

The magnitude scale originated with the Greek astronomer Hipparchus in the second century B.C. He simply divided the stars into five groups of what he considered to be similar brightness. All stars were classified as first, second, third, fourth, or fifth magnitude. Later, Ptolemy adopted Hipparchus's system, and it became the standard system of stellar brightness of the ancient and medieval worlds. Among the many things that Galileo discovered with his telescope was the existence of stars too faint to be seen with the naked eye. This required that the system be extended to higher magnitudes to include these stars.

The modern magnitude scale was created in the 1850s by Norman Pogson, using logarithmic measures of the celestial objects' brightnesses. The modern magnitude system is defined such that a factor of ten in brightness corresponds to a difference of 2.512 magnitudes. In this system, the brighter a star is, the smaller the value

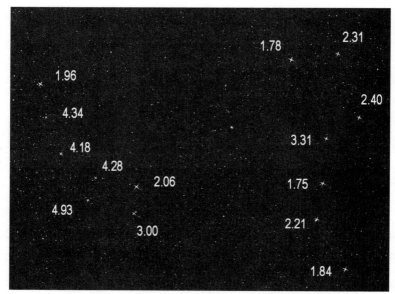

The Little Dipper (left) *and Big Dipper* (right) *are perfect for showing differences in stellar magnitude.* ILLUSTRATION BY DAVID FRANTZ; PHOTO BY VIC WINTER

of its magnitude. In fact, the brightest stars have negative magnitudes. This means a star of magnitude 1 is 2.512 times brighter than a star of magnitude 2. Since this is a logarithmic change, you multiply brightnesses for each added magnitude. A magnitude 1 star is 6.310 (2.512 × 2.512) times brighter than a magnitude 3 star, or 100 times brighter than a magnitude 6 star.

There are other terms you will come across when exploring the night sky. Positions are given in declination (DEC) and right ascension (RA), coordinates that are much like Earth's latitude and longitude system. The celestial coordinate that corresponds to latitude, declination, is measured in angular units of degrees, minutes, and seconds, going from +90 degrees at the North Celestial Pole (NCP) to –90 degrees at the South Celestial Pole (SCP). These minutes and seconds are actually arc minutes and arc seconds. The sizes of objects in the sky are also given in degrees, minutes, and seconds, here again arc minutes and arc seconds. For example, the moon and sun are approximately 30 arc minutes in diameter as

Celestial objects are located by stellar longitude and latitude, or right ascension and declination. Right ascension is shown here, along with constellation outlines. PHOTO BY RYAN DIDUCK AND MIKE REYNOLDS

seen from Earth. Like Earth's equator at 0 degrees latitude, the Celestial Equator is at 0 degrees DEC.

The celestial coordinate that corresponds to longitude, right ascension, is measured in hours, minutes, and seconds. (An arc minute is ¹⁄₁₅ the size of a minute of right ascension, and an arc second is ¹⁄₁₅ the size of a second of right ascension.) Right ascension goes from 0 to 24 hours, with each hour of RA representing 15 degrees, and the full 24 hours of RA representing 360 degrees. This is a sensible system, as the celestial sphere is like a big clock in the sky, turning as Earth rotates. In fact, before the 1950s and the advent of atomic clocks, the sky was the world's time source.

STAR COLORS

Even with the unaided eye, you can see that stars do indeed have colors. These colors basically are indicative of the star's temperature. Colors range from the cooler red stars, through orange, yellow, and white, to blue for the hottest stars. A star's color tells astronomers a lot about its age and size. In basic astronomy textbooks, you will often see a graph of stars plotted by their temperature (or color) versus their absolute magnitude (or luminosity). Such a graph is called the Hertzsprung-Russell Diagram.

Stars are classified according to their temperatures, and thus colors. Our sun, a yellow star, has a primary classification as a G star. Cooler red and orange stars, such as Antares and Arcturus, are classified respectively as M and K stars, and hotter yellow-white, white, and blue stars, such as Procyon, Sirius, and Rigel, as F, A, and B stars. There are a few very hot stars, such as the two end stars of Orion's Belt, Alnitak and Mintaka, that are classified as O stars.

DOUBLE AND MULTIPLE STAR SYSTEMS

One of the joys of binocular stargazing is viewing double star systems. Star colors are especially noticeable when two stars of similar brightness but different colors are seen side by side. Imagine a system with a star revolving around a central sun instead of, or in addition to, planets. Physically associated systems made up of two stars are called binary stars. The two stars are in orbit around the system's center of gravity, which depends on the mass of each star. Occasionally two stars are not physically related to each other but just happen to line up so that they appear to be a double star system. Such a case is called an optical double. Often the unaided eye cannot separate, or split, a pair of double stars, meaning that you cannot see both stars individually. You also may come across multiple star systems, containing more than two stars.

The distance between component stars is referred to as the separation, measured in arc minutes or arc seconds. Also measured is the position angle, which refers to an imaginary line going from the primary star (the brighter star) to the secondary star relative to north. If the position angle (PA) of the two stars is 0 degrees, the secondary is due north of the primary. At a PA of 90 degrees, the secondary star is due east of the primary star.

There is often a beautiful contrast in the colors of the double or multiple systems' component stars. For example, the famous Alberio, or b Cygni, is a bright double star system with gold and sapphire stars. The stars are magnitude 3.1 and 5.1 objects. At a separation of 34 arc seconds and a position angle of 54 degrees, it is an easy double star to split with medium-power binoculars.

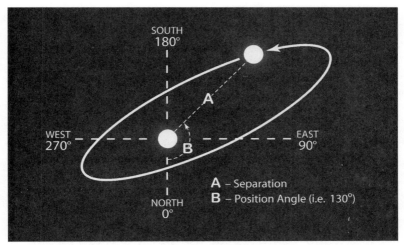

The relationship between double stars is shown as the separation A, in arc seconds, between the primary and secondary stars, and the position angle (PA) B from north at 0°. In this example, the PA is approximately 145°. ILLUSTRATION BY DAVID FRANTZ

Then there's the famous "double-double": Epsilon Lyrae, or e Lyrae. From our vantage point, the two main stars are 208 arc seconds apart, and some observers with excellent eyesight can split the stars without binoculars or a telescope; however, a good telescope at high magnification must be used to separate each component into close binaries of 2.7 and 2.3 arc seconds.

Another well-known double is Alcor and Mizar in Ursa Major. Even though these stars can be split with the unaided eye, they make a nice binocular pair. This pair was first recognized by the Italian astronomer Giovanni Ricciolo in 1650.

Part of the challenge with double stars is to see what you can split with your unaided eyes and with binoculars. A steady mount is necessary for some close doubles; vibrations while hand-holding will cause you to see multiple stars—but not the type you are trying to observe. Generally speaking, the higher the magnification of the binoculars, the closer the stars you can split. Do a test to determine the smallest separation your binoculars are capable of. A rule of thumb is that you can split a double separated by about 40 to 45 arc seconds with 7-power binoculars (this rules out Alberio for low-power binoculars), 25 to 27 arc seconds with 11-power, and 12 arc seconds with 25-power. Often one of the double or multiple system's components is faint, making it even more challenging to split the stars. The closer they are in magnitude, the more likely the brighter component will overcome the other star in the pair.

The universe is not a static place, and physical doubles can change positions. A table or book might list the separation and position angle of double stars, but this will eventually change. Sometimes this happens very slowly and is barely noticeable, but other times it is faster, with recognizable changes in brightness.

ECLIPSING BINARY STARS

Imagine a double star system whose orbital plane is seen edge-on from our viewing perspective. The component stars, while orbiting their center of mass, periodically eclipse each other. What would we see from our vantage point? Most likely we would see what appears to be one star that is steady in brightness, then fades, depending on the position of the two stars to each other. These are called eclipsing binary stars (EBSes). The period is generally measured from the center of one primary eclipse until the next, and it can range from minutes to years. Most well-known EBSes have periods of a few days to a few weeks.

The two component stars usually are not similar. The size and brightness of both make a difference in the minima—the star's dimmest brightness. The deeper eclipse is called the primary eclipse, or minimum, and the shallower eclipse is called the secondary eclipse. The eclipse can be central, with a flat bottom to its light curve, or partial, with a V-shaped light curve. A light curve is a graph of a star's magnitude versus the time of each magnitude estimate. The duration of the minimum—the period in which the combined light of the stars is dimmed—can be anywhere from a few hours to a couple days for well-known EBSes. At the extreme end is epsilon Aurigae, which has a period of 27 years and minima durations of about 1.5 years. It will next enter eclipse in 2009.

Probably the most famous and one of the most frequently observed eclipsing binaries is Algol. Algol has appeared to wink as it varied over a short period of time, inspiring myths that it was a winking demon's eye. Homer wrote of Algol in the *Iliad,* "the Gorgon's head, a ghastly sight, deformed and dreadful, and a sight of woe." Its name is Arabic for "Demon" or "Demon's Head." It has also been called the Demon Star, Satan's Head, Ghoul, and by the Chinese, Tseih She, meaning "piled-up corpses."

This EBS is an easy unaided-eye object with a period of less than 3 days, going from magnitude 2 to 3, and then back to magnitude 2 again, repeating the cycle. The discovery of Algol as a variable is credited to the Italian astronomer Geminiano Montanari in 1667. It turns out that Algol is actually a multiple star system with three stars. The third star is quite faint and orbits the other two stars every 1.86 years. John Goodricke, an English astronomer, determined Algol's periodicity in 1782–83.

Binoculars are useful tools for observing many of the EBSes. The American Association of Variable Star Observers (AAVSO) has excellent programs for observers interested in EBS work. You will observe the EBS for a period of time around the predicted minima, estimating the brightness of the EBS by comparing it with other stars. A graph of the EBS's magnitude versus the time of each of your estimates produces the EBS's light curve.

The key to contributing to programs like these is to be well prepared. Make certain you have the necessary data, which includes the predicted minima as well as the duration of the eclipse itself; the AAVSO has its *Bulletin* available. For many of these EBSes, charts are available that show the location not only of the EBS, but also of stars with which to compare the EBS to estimate its brightness while graphing the descending and ascending branches of its minima. Observation about once every ten minutes is usually called for, with observations timed to the nearest minute. For this, an accurate watch or clock is necessary, especially in timing the EBS's minima. Record what you see, not what you think you should see according to the predicted minima.

VARIABLE STARS

Variable stars are stars that appear to change in brightness. Some of them change in brightness like clockwork; others are irregular and unpredictable. The stars vary in brightness for different reasons, depending on the star itself.

Observing variable stars is somewhat like observing eclipsing binaries. You need an atlas to locate the general region of the sky where the object is located. If you want to make serious observa-

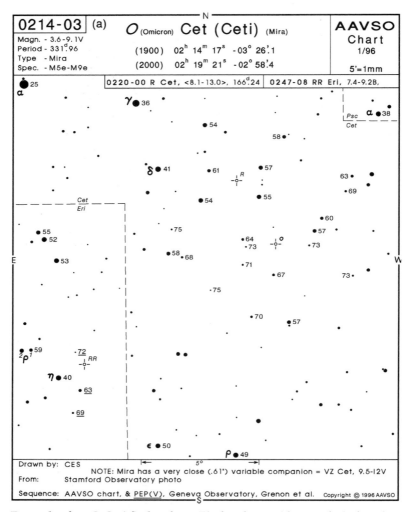

Example of an O Ceti finder chart. Finder charts aid not only in locating an object but in estimating its change in magnitude over time; they also provide other pertinent information about an object. COURTESY OF AAVSO

tions of variable stars, contact the AAVSO. Membership provides you with significant expertise and guidance to hone your observing skills. The AAVSO publishes charts that not only assist in locating the variables, but also indicate comparison stars for estimating the brightness of the variables. It is helpful and really a

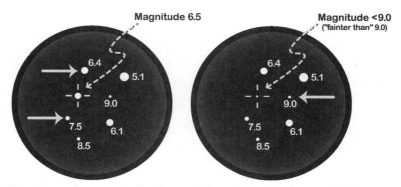

To estimate the magnitude of a variable or EBS, the observer looks at nearby stars and interpolates the star's brightness. On the left-hand side, the object's interpolated magnitude is 6.5, a little fainter than the 6.4 star but much brighter than the 7.5 star. The example on the right shows a star fainter than the faintest star on the chart, interpolated to be around magnitude 9.0. ILLUSTRATION BY DAVID FRANTZ

must to have identified these stars before making estimates. Star-hopping with binoculars, using a planisphere and star chart, takes some practice for new observers. Look for asterisms, such as a line or triangle of stars, that will point you in the right direction or help you locate the specific star.

As with eclipsing binaries, predictions of minima for variable stars are available from the AAVSO. If you decide you want to determine the times of minimum light for these stars, you need to compare the variable with the comparison stars at specific time intervals. For cataclysmic variables, these time intervals are usually every 10 to 15 minutes, unless otherwise noted. Look at the variable and determine which of the comparison stars is closest to it in brightness. At most times, the variable star will be between two of the comparison stars in brightness. In order to estimate the magnitude of the variable, it is necessary to estimate (linearly) where the variable falls on the intensity range represented by the difference in brightness between the brighter and fainter comparison stars. This process is called interpolation. Carefully record your estimate and the time it was made. For longer-period variables, the interval between observations may be a week or more.

With supernovae and novae, you hope to catch an event in progress. How exciting it would be to see some of these events—

an incredible brightening of many magnitudes. For many of us, a bright nova or supernova is a once-in-a-lifetime happening.

Variable stars can be extrinsic or intrinsic. Eclipsing variable stars are referred to as extrinsic variables; the variations in light that we see are due to the eclipse of one star. Intrinsic variable stars vary in brightness as a result of physical changes in the star or the stellar system itself. There are a number of types of intrinsic variables, which are divided into two main groups: pulsating variables and cataclysmic variables.

Pulsating Variables

Pulsating variables are stars that undergo a periodic expansion and contraction of the surface. There are five general classes of pulsating variables, listed in order of light variations.

Cepheids. Massive, high-luminosity yellow stars with short periods of 1 to 70 days and light variations of 0.1 to 2 magnitudes.

RR Lyrae. White giant stars with short periods of 0.5 to 1.2 days and light variations of 0.3 to 2 magnitudes.

RV Tauri. Yellow supergiant stars experiencing alternating shallow and deep minima, with periods of 30 to 150 days between the deep minima and light variations up to 2 magnitudes.

Long-period. Pulsating red giant or supergiant stars with periods of 30 to 1,000 days. These include the Mira variables, named after the star Mira (also known as Omicron Ceti), with light variations of more than 2.5 magnitudes, and semiregular variables, which experience periodicity as well as intervals of semiregular or even irregular light variations of less than 2.5 magnitudes.

Mira's name means "the amazing one." Mira is the only named star in the sky that, for a period, is too faint to be seen with the naked eye. Mira is the brightest of the long-period variables, thousands of which are now known. Mira is one of the coolest stars in the sky, with a temperature just above 2,000 degrees Kelvin. However, this star is approaching the last stages of its life. The light variations are caused by pulsations, changes in size that also affect temperature. Eventually Mira's outer stellar layer will be blown away and all that will remain will be a white dwarf the size of the Earth. Many billions of years from now, the same thing will happen to our sun.

Irregular. Stars that demonstrate changes with no discernible period. These include the majority of red giants.

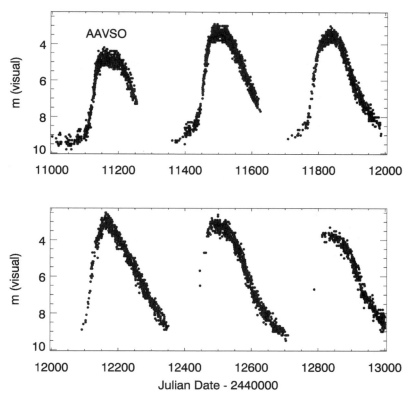

Shown here are some examples of O Ceti light curves. By taking observers' data, one can plot the changes in the magnitude (m) over time. COURTESY OF AAVSO

Cataclysmic Variables

Cataclysmic variables, also referred to as eruptive variables, are stars that have extreme outbursts because of the stellar processes within the interior or on the surface. There are six general classes of cataclysmic variables.

Supernovae. A supernova is an event in which a massive star undergoes a sudden cataclysmic stellar explosion that results in a brightness increase of 20 magnitudes or greater. There are two types of supernovae. A type I supernova is caused by infalling matter on a white dwarf star and does not have hydrogen lines in its spectra. (A spectrum—plural spectra—is seen when light is broken or split into its component colors.) A type II supernova occurs when a massive star of at least three solar masses (that is, three times the mass of the sun) explodes; it has hydrogen lines in its spectra.

Novae. A nova is a dramatic brightening of a white dwarf primary star that is accreting (a process whereby orbiting material builds up) in a close binary star system when the shell surface on the primary star undergoes a burst of hydrogen fusion. The system can brighten 7 to 16 magnitudes over one to several hundred days; fading may take decades.

Recurrent novae. A recurrent nova is similar to a nova but experiences two or more somewhat reduced outbursts.

Dwarf novae. A dwarf nova occurs in a close binary system composed of an accretion disk (material that has built up around a star) surrounding a white dwarf and a red dwarf, when the accretion disk is unstable and brightenings of 2 to 6 magnitudes occur.

Symbiotic stars. These are close binary systems composed of a red giant and a hot blue star, both of which are embedded in nebulosity. Semiperiodic outbursts of up to 3 magnitudes occur.

R Coronae Borealis. These are rare supergiants that, after a prolonged period of time at maximum brightness, occasionally fade as much as 9 magnitudes at irregular intervals, returning to maximum brightness after a few months to a year.

7

Deep-Sky Objects

Deep-sky objects are extended objects beyond our solar system. (Occasionally you will hear solar system objects referred to as shallow-sky objects.) "Extended" refers to the fact that from our vantage point, the object has length and width rather than appearing as a point of light like a star. Observing deep-sky objects, unlike the sun, moon, and planets, is best done under dark, starry skies. Light pollution from streetlights, billboards, and general urban lighting interferes with deep-sky observing, although a number of objects still can be seen in a light-polluted setting. A light-polluted sky creates a bright background and causes these dim extended objects to fade into the light fog.

Deep-sky objects (DSOs) include star clusters, gaseous nebulae, and galaxies beyond our own Milky Way. Even though they are grouped together as deep-sky objects, each type is quite different in nature, in both composition and appearance.

STAR CLUSTERS

Star clusters are groupings of stars that are categorized as open or globular. Open clusters have fewer stars than globular clusters, and they are found in different locations. Globulars are generally very old, some of the oldest classes of objects, whereas open clusters are composed of very young stars. These objects are crucial in our understanding of the age of the universe.

Open Clusters

Open clusters are groups of about a dozen to a few thousand stars that are physically related to one another by being held together by their mutual gravity. Many of the brightest—and prettiest—open clusters contain very luminous, young, hot blue stars or the red supergiants that these blue stars eventually become. These stars stand out against the background of the rest of the cluster members. Believed to originate from the Milky Way's gas and dust clouds known as diffuse nebulae, open clusters tend to take up a rather limited area and orbit in the main plane of the galaxy's disk. Stars within open clusters are similar because they appear to be of approximately the same age and chemical makeup.

More than 1,000 open clusters have been identified in the Milky Way, though some astronomers speculate that there could be more than 100,000 open clusters within our galaxy. Some of the largest open clusters can be seen in nearby galaxies, such as M31 and M33. The life spans of open clusters are relatively short, as member stars begin to escape the cluster as a result of galactic interactions with other clusters, stars, nebulae, and the galactic nucleus.

Open clusters are often described by several factors, such as the number of stars within a specific cluster. Poor clusters have less than 50 stars, moderately rich have 50 to 100, and rich have

Even thirty miles from a large city, light pollution can be a problem. This five-minute exposure shows star trails and, at the tree line, the lights of Jacksonville, Florida.

The Pleiades, M45, one of the finest open star clusters in the heavens
PHOTO BY VIC WINTER

more than 100. Other descriptors include the degree of concentration and range of brightness of the stars within a specific cluster.

Some open clusters have been known about since prehistoric times, including the Pleiades (M45), Hyades (actually made up of older stars), and Beehive, or Praesepe (M44). As early as 138 A.D., Ptolemy mentioned M7 and the Coma Star Cluster. Galileo observed M44 and noted its stellar composition.

Many open clusters are wonderful subjects for binoculars. Open clusters such as the Pleiades look their best when observed at a very low power and with a wide field of view. Many open clusters are reminiscent of blue-white diamonds sprinkled in a group. Some, like M67, look quite different because they are composed of red to yellow stars; all of the blue stars have long since exploded. Even fainter open clusters are quite spectacular, but these can pose a challenge when using a pair of binoculars.

Globular Clusters

Globular clusters are also gravitationally bound groupings of stars but differ from open clusters in a number of ways. First globulars, as they are often called, are made up of tens of thousands to more than a million stars (Omega Centauri may have more than 5 million). Globular clusters populate the halo around the galactic center, or bulge, of our Milky Way and other galaxies, with a strong concentration toward the galactic center.

Globular clusters are quite different in appearance from open clusters. As the term "globular" implies, these objects are spherical and clumplike in appearance. Some globular clusters seem to be more centrally condensed, with a thinner halo of stars around

M13, *the beautiful globular cluster found in the constellation Hercules* PHOTO BY VIC WINTER

them; others are almost uniform in brightness across their surface. Higher powers of magnification and larger light-gathering objectives, along with a steady mounting point, are needed to see these kinds of details with binoculars.

Many globulars can be a challenge to find and observe through binoculars or small telescopes. There are exceptions, however. Ptolemy first noted the spectacular globular cluster Omega Centauri, which is visible to the naked eye. The best-known globular cluster in the Northern Hemisphere is M13 in Hercules, and several excellent southern globular clusters in the Scorpius-Sagittarius region, such as M4, are visible from most of the United States.

GASEOUS NEBULAE

Nebulae represent a number of different types of objects that consist of clouds of dust and gas. The word *nebula* has had various meanings, but in its original Latin, it meant "cloud." With low-power optics with a small aperture (2 to 4 inches in diameter), many deep-sky objects—star clusters, gaseous nebulae, and external galaxies—appear to be small patches of cloud against the night sky. Before the invention of the telescope, nebula was used to denote an extended object versus a point of light (star). These objects would have included open star clusters such as the Pleiades. After the invention of the telescope, other deep-sky objects, including galaxies, were also referred to as nebulae. Astronomy books from the 1800s, for instance, make reference to the Andromeda Nebula.

Today nebulae refers only to interstellar clouds of dust and gas, not clusters of stars or external galaxies. Gaseous nebulae are categorized into different types based on the light we see or how they are formed. These groups include diffuse, emission, reflection, and dark, or absorption, nebulae. A nebulae can fall into more than one category.

Diffuse nebulae are clouds of interstellar matter: gas and dust. Star formation can take place within a diffuse nebula if it is big and massive, and thus clusters of stars can be born; however, once star formation starts, the diffuse nebula rapidly becomes an emission nebula. Diffuse nebulae are often seen because they partially block starlight or reflect it. There are three types of diffuse nebula: emission, reflection, and dark.

An emission nebula is formed when hot, energetic stars within a diffuse nebula cause the nebula's gas to become "excited" and glow. The Great Orion Nebula, M42, is a classic example of an emission nebula. Though smaller binoculars, it appears gray, but with midsize optics, such as 6-inch binoculars or an 8-inch telescope, it looks faintly green. This is because of the eye's sensitivity to the green part of the spectrum. Only with relatively large telescopes, say 16 inches or larger, is enough light collected to trigger the eye's red color vision and make this nebula appear red. Because modern photographic film and charge-coupled device (CCD) detectors are much more sensitive than the human eye to red light, emission nebulae appear red in color photographs and CCD images.

If the nebula is visible as a result of starlight reflected off its dust, it is considered a reflection nebula. Many emission nebulae contain dust and thus are also reflection nebulae. One example is the Flame Nebula, NGC 2024, next to Zeta Orionis (often seen in the same field with the Horsehead Nebula in photographs). The

M42, the Great Orion Nebula. This is one of the finest deep-sky objects seen from the Earth; it is quite stunning when viewed through binoculars. PHOTO BY VIC WINTER

reflected portion of the light from a nebula such as this one appears to be blue, since blue light is much more strongly scattered than red.

Unlike emission or reflection nebulae, dark nebulae, also called absorption nebulae, are visible because they totally absorb starlight from objects behind them so that none of it reaches us as observers. Dark nebulae are made up of dust and gas, but they have no stellar illumination. The Horsehead Nebula is one example. The stars illuminating the background nebula are hidden behind the dark nebula, which by chance forms the horse's head.

Emission, reflection, and dark nebulae are usually star-forming, or prestellar, nebulae and may involve many stars. (In many star-forming regions, no stars are visible because they are all buried deeply with the dusty gas; however, in the near infrared, they appear to be lit up like Christmas light displays, because infrared light can escape the nebulae.) These differ from the post-stellar nebulae—the planetary nebulae and supernova remnants—all of which involve one specific star.

The Horsehead Nebula, an example of a dark nebula. It is caused by dust that blocks starlight. PHOTO BY JERRY ARMSTRONG

The Ring Nebula, M57. This is an excellent example of a planetary nebula. PHOTO BY JERRY ARMSTRONG

Though technically they are also gaseous nebulae, planetary nebulae and supernova remnants are always referred to by those names and are not included in the category of "gaseous nebulae." This is because they derive their appearance and generate their emission by driving a shock wave into a cloud of gas that they have ejected, either slowly during their red giant phase or rapidly during an explosive event such as a nova or supernova explosion.

Planetary nebulae are formed when a star similar to our sun, and up to eight times more massive, has used up all of its nuclear fuel, ejecting a significant percentage of the star's mass in a shell. (There is currently debate over whether our sun's remnant will be energetic enough to excite a planetary nebula into existence at the end of its nuclear fuel burning lifetime.) This shell of gas is visible because it is being excited by the high energy of the hot star that ejected it. The shell expands and eventually fades. Observationally, a planetary nebula can look like a smoke ring or even the rings of Saturn.

Supernova remnants, expanding clouds of gas and dust, become visible when stars roughly over eight times more massive than our sun explode, in what is known as a supernova.

EXTERNAL GALAXIES

Galaxies are large systems of stars and matter, usually also containing various types of star clusters and nebulae. Those that can be seen with binoculars and small telescopes typically contain hundreds of millions to trillions of stars. The sizes of galaxies can range from a few thousand to several hundred thousand light-years. And galaxies are usually separated by distances of millions of light-years.

Our home galaxy, the Milky Way, is a spiral galaxy about 100,000 light-years in diameter. It has an estimated 100 billion stars (although I've heard a number as high as 400 billion) with a mass of about a trillion suns. We see a cloudlike band of light overhead in the night sky during both summer (stronger) and winter (weaker) months, which we call the Milky Way, though this is simply where the Milky Way is the brightest. (During spring and fall months, we are looking out of the plane of the galaxy toward intergalactic space.) Most objects we see with our naked eye in the night sky are Milky Way objects. Optical aid allows us to see other galaxies, however. They come in a variety of shapes and types, including elliptical, spiral, barred spiral (with a bar-shaped central bulge), lenticular, and irregular.

The Andromeda Galaxy is a beautiful and possibly larger spiral galaxy about 2.1 to 2.5 million light-years away. It has a pair of dwarf elliptical companion galaxies, M32 and NGC 205, which are

The Andromeda Galaxy, M31, one of the most striking objects you can see through binoculars. An elliptical galaxy companion, M32, can be seen above and to the right. PHOTO BY JERRY ARMSTRONG

visible in binoculars. Other galaxies can also be seen, such as the irregular Large and Small Magellanic Clouds.

CATALOGING DEEP-SKY OBJECTS

Because thousands upon thousands of deep-sky objects have been identified, it's important to have a system to catalog and identify the type of object, its location in the sky, magnitude, and other traits. A number of cataloging systems are in use today.

Charles Messier, a French astronomer and comet hunter, compiled a list of deep sky objects while searching for comets that still bears his name. The Messier Objects, or M Objects, are some of the brighter objects visible from the Northern Hemisphere. For example M42, or Messier 42, is the Great Orion Nebula.

Messier started his cataloging while searching for comets because, with the telescopes of his day, deep-sky objects resembled comets in every respect except for motion against the sky. The list of Messier Objects is a Northern Hemisphere catalog consisting of 110 numbered objects, of which 109 can be located, but it is not all-inclusive (for example, the spectacular Double Cluster in Perseus and several dark nebulae are conspicuously missing from the list). Nevertheless, it makes a good binocular list to start with.

J. L. E. Dreyer's *New General Catalog* (NGC), first published in 1888, is more complete than Messier's catalog, and it includes the Southern Hemisphere objects. There also are other catalogs, generally named after their creators, such as the Herschel and the Caldwell Objects lists.

Some objects are known by a popular name as well as a catalog number, such as the Great Orion Nebula, which is also M42.

FINDING AND OBSERVING DEEP-SKY OBJECTS

There are many techniques people use to observe deep-sky objects, depending on the object and the type of instrument being used. Binoculars, with their wide field of view, allow you to star-hop, going from one familiar object or star to the next object in a line to the object you wish to observe. To do this, you need to learn the sky and have a good star atlas at hand. By star-hopping, you not only will learn more, but also are likely to see many more objects as you make your way to the object that is your goal. And to be honest, it is a lot of fun. Another method is to simply scan the sky and then try to identify something that appears interesting, referring to your star atlas. This can be enjoyable as well.

Before beginning your observation, make certain your eyes are dark adapted. Spend some time out of light, allowing the pupils to dilate to their fullest so that more light can enter. This process may take as long as thirty minutes and seems to take longer in older observers. Light of any type, including moonlight, can interfere with dark adaptation. Use a red-filtered flashlight to view your star atlas.

Don't simply glance at an object and then look away from it. Spend time exploring the object; do not be impatient. Look for such things as differences in star magnitudes and colors. Note subtle changes in nebulosity. If the object is a globular cluster, see if you can resolve stars in the cluster; how does the center appear to you? Can you see color? In many objects, color is visible only with much larger instruments, so don't be disappointed if you do not see the same colors as in photographs (emulsions and imaging devices build up an image over time, and processing techniques may be used to enhance color in professional images).

Many observers use a technique called averted vision, which means they do not look directly at the object, but a little off to one side. The sensitive part of the eyes is the rods, which are not in the center (the cones are). By looking off to the side, the rods play a bigger part, thus allowing you to see fainter objects.

DRAWING DEEP-SKY OBJECTS

Drawing deep-sky objects is a real gift. Dave Branchett, whose drawings of deep-sky objects appear in this book, has been drawing objects for many years. He recommends starting with a circle on the drawing paper to represent the field of view, which you then place over a star chart. Begin by recording the more prominent stars from the chart that you find in your field of view, which sets the foundation for the drawing. Then, looking through your binoculars, carefully and slowly sketch the details of the object: nebulosity, fainter stars, and so on.

Like Dave, Gonzalo Vargas has also been drawing deep-sky objects for many years. His perspective has been from Bolivia and as a Southern Hemisphere observer. And Gonzalo's techniques are a little different than Dave's. First he develops an observational

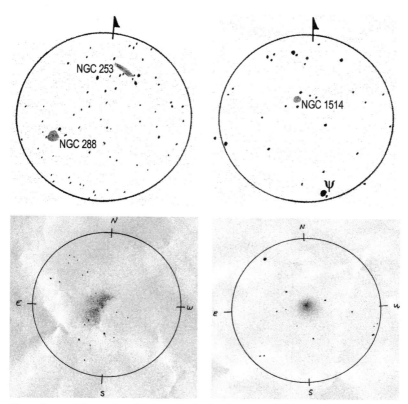

Drawings of deep-sky objects, clockwise from top left: NGC 253 and NGC 1514; 47 Tucanae; M42 TOP TWO DRAWINGS BY DAVE BRANCHETT; BOTTOM TWO BY GONZALO VARGAS

plan, consults star charts, prepares his drawing paper with a circle to represent the field of view through the binoculars (or telescope). Then Gonzalo allows his eyes to adapt to the dark for at least 15 minutes. When he begins to observe, he looks at all the details in the eyepiece, including the target object and bright stars. The drawing begins with stars, sketching geometric configurations using a regular pencil for the initial sketch. He has learned to use both eyes, one at the telescope and one on the drawing, going back and forth between the two. After drawing bright and fainter stars, he then begins to draw the actual object itself, looking for brighter and darker zones. Gonzalo uses a soft pencil for nebulae and will actually add detail by smearing with his finger. He then looks to see if the drawing is accurate, makes corrections as needed, and adds details like north and south.

Dave notes that binocular objects, for the most part, are small, faint, and opaque and thus are relatively easy to record. All of his sketches in this book were made while using 16×80 binoculars with a 3-degree field of view. As you continue to develop your techniques, you will learn what is pleasing to your eye in the final drawing. Dave notes that his techniques have sharpened over the years.

Take your time and record your observations as accurately as possible. A drawing of M35 took Branchett three hours to complete. He likes listening to music as he observes; in particular, he recommends Mozart. By drawing what you see, you learn to improve your observation and become more intimately acquainted with these spectacular objects.

8

Spring Objects

Before we begin, a few notes about this section of the book: This chapter and the following three look at the skies during each of the four Northern Hemisphere seasons. They are not all-inclusive lists or star charts; there are numerous references, star charts, and atlases you can consult for this purpose. The goal of these chapters is to start the beginning binocular observer on his or her way. They briefly describe some of the best objects and what you might expect as you hunt for them.

Many books do not attempt to divide the heavens into four seasons for several reasons. First, strictly dividing the heavens into four quadrants results in several constellations that lie across two seasons, and they have to be assigned to one or the other. Second, as the night begins, depending on the time of the season during which you are observing (beginning, middle, or end), constellations from the previous season are still visible in the west. And as the night passes, constellations from the next season begin to rise in the east. I tried to set these seasonal chapters for the beginning of each season at about 9:00 P.M. But even then, the sun has not set for many during the summer (also a product of daylight saving time). In spite of these drawbacks, I felt it best through my experience to provide you with a by-season observing guide.

Each chapter sets a rhythm for its seasonal overview. First the constellations of the season are noted, including what I call a "low in the south" grouping—constellations that need to be observed from the southern latitudes to have the best view, if any, of objects within. Even then, they can be a challenge to observe. Many of the fabulous "low in the south" objects are discussed in chapter 12, which is specifically on the southern skies.

Next a number of the objects are highlighted, followed by tables listing an overall summary of many of the top binocular objects for the season, occasionally challenging. These are broken into three major areas: deep-sky objects, stars, and variables.

The deep-sky objects—globular clusters, open clusters, nebulae, and galaxies—are listed by object name (M = Messier; C = Caldwell; NGC = New General Catalog) along with other names, if any, followed by the type of object, constellation in which it can be found, brightness, size in minutes of arc, and comments as appropriate. These objects should all be visible through 50-millimeter binoculars unless otherwise noted. Keep in mind that the moon is about 30 minutes of arc as you consider the size of these objects.

As you review various deep-sky object lists and star charts, you might note discrepancies in size and magnitude. Sometimes the magnitude is a visual measurement, and other times it is photographic. Also, a seventh-magnitude galaxy that is small appears brighter than a sixth-magnitude object that is more spread out. In both my own observations and various other sources, I have seen variations of up to 3 magnitudes on the same object. The sizes given can be larger when measured photographically, because film and digital pick up light over a period of time, best described as an additive process. In preparing this text, I tried to use the most consistent data, coupled with my personal observing experience.

The stars listed are some of the more extraordinary stellar objects, remarkable for their brightness or color, or because they are double stars. These are great objects with which to start your tour of our stellar neighborhood.

The variables include some of the finest eclipsing binary and variable stars of each season. Variables and EBSes are designated by a letter, the first variable named in a specific constellation beginning with an uppercase letter, then the three-letter constellation abbreviation as determined by the International Astronomical Union. For example, S Antliae, in the constellation Antlia, is designated S Ant.

How often should you observe a variable? That depends on the type of object. The American Association of Variable Star Observers has set guidelines for these and other types of variables: Cepheids should be observed every clear night; irregular, Mira-type long-period, and semiregular variables once a week; and eclipsing binaries every ten minutes during an eclipse.

The Astronomical League has an outstanding series of programs for observers. You also need some star charts in addition to a planisphere. There are a number of excellent sources available. One of my favorites for the Messier Objects is the late Harvard Pennington's *Year-Round Messier Marathon Field Guide.* Online sources include David Paul Green's Ultimate Messier Object List, Overlooked Object Log, and Caldwell Observer's Log. Green's shareware provides you with a page-by-page reference, including a star chart and other information, a log to record your observations, and sketching forms.

Spring brings about warmer and—in many locations—better weather. Stargazers begin to come out of their winter hibernation, dust off their binoculars or telescope, and prepare to assault the skies. The spring skies offers the binocular stargazer many wonderful objects with which to shake off that winter chill.

One annual springtime ritual is an event called the Messier Marathon, when observers try to see nearly all of the Messier objects one night. This occurs around mid- to late March. Because of the celestial distribution of the Messier objects, people at sites farther south have a better chance of observing all the objects. You have to be fast and well organized, especially at the start of the evening, as several objects in the west are quickly lost as they begin to get low on the horizon. No moon or very near new moon is essential to successfully completing the Messier Marathon.

The idea of a Messier Marathon was first suggested in the 1970s. Over the years, it has taken on a variety of forms, most recently the M^3, or M-cubed—Messier Memory Marathon—in which observers find the objects by memory without using any star charts or notes. Some observers complain that the marathon concept does not allow them to enjoy and study the objects. But you can always spend more time viewing them later. There are a number of good websites and books on the subject.

For novice observers, the springtime night skies offer some bright constellations (defined as official groupings of stars) and asterisms (unofficial groupings) to guide the way. Brighter constellations that can help you find objects in the fainter spring constellations are Leo, the Lion, and its easily recognizable sickle; Ursa Major, the Big Bear (many people better recognize this as its asterism, the Big Dipper); and Bootes, the Herdsman. You can also look for Antlia, the Air Pump; Canes Venatici, the Hunting Dogs; Coma Berenices, the Hair of Berenice, wife of Egyptian King Ptolemy III; Hydra, the Water Snake, a constellation that literally snakes across the sky; Leo Minor, the Small Lion; Libra, the Scales; Sextans, the Sextant; and Virgo, the Virgin. Three of these constellations, Leo, Libra, and Virgo, are found in the zodiac. Low in the south, you will find Carina, the Ship's Keel; Centaurus, the Centaur; Corvus, the Crow; Crater, the Cup; Crux, the Southern Cross; and Vela, the Ship's Sails.

DEEP-SKY OBJECTS

M3. A magnitude 5.9 globular cluster in Canes Venatici; look for an object that is a little bigger than 16 arc minutes in diameter. One of our brightest Northern Hemisphere globular clusters and easy in 7×50s.

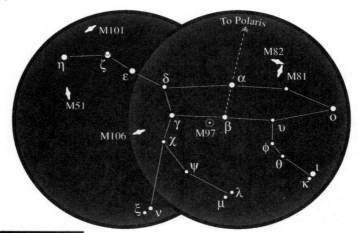

- ● Star
- ◥ Galaxy
- ◉ Globular Cluster
- ◉ Open Cluster
- ▬ Nebulosity
- ⊙ Planetary Nebula

Shown here are Ursa Major—the Big Bear—and several objects, including a number of galaxies, that lie within its boundaries. One circle is approximately 35 degrees; the distance between a Ursa Majoris and b Ursa Majoris is a little over 5 degrees. ILLUSTRATION BY DAVID FRANTZ

M51, the Whirlpool Galaxy
PHOTO BY JERRY ARMSTRONG

M51. The Whirlpool Galaxy, a spectacular example of a spiral galaxy in Canes Venatici as a "bird's-eye" view at magnitude 8.4. About 10 by 6 arc minutes in size.

M63. A spiral galaxy in Canes Venatici at magnitude 8.6. You will find this 10 by 6 arc minute galaxy a bit of a challenge in 7×50s.

M94. Another spiral galaxy also in Canes Venatici at magnitude 8.2. You will find this 7 by 3 arc minute galaxy a little more challenging than M63.

Melotte 111. Better known as the Coma Berenices Star Cluster, this nice open cluster of about sixty stars, located in Coma Berenices, has a handful of magnitude 5, 6, and 7 stars—about eighteen total. Melotte 111 is fairly large, at about 5 degrees in size, and simply a terrific binocular object.

M48. A nice magnitude 5.8 open cluster located in Hydra that is about 54 arc minutes in size. An easy object in 7×50s.

NGC 3242/C59. Also known as the Ghost of Jupiter, this planetary nebula in Hydra is magnitude 7.8 and about 0.3 arc minutes in size.

M104. Also known as the Sombrero Galaxy, this 50-millimeter spiral galaxy found in Virgo is at magnitude 8.0. It is also rather small, at 9 by 4 arc minutes, and very challenging for binoculars. You might note a large number of galaxies in Virgo, as well as Leo. Many of these are very faint, requiring giant binoculars or, more often, a telescope. As a challenge, see what is the faintest galaxy you can observe at a dark-sky site with your binoculars in this galaxy-rich region of the sky.

*M104, the Sombrero
Galaxy* PHOTO BY VIC WINTER

STARS

Alpha Bootis. More familiar to amateurs as Arcturus, this brilliant magnitude –0.06 star with an orange color is a nice binocular object.

Alpha Leonis. The brightest star in the constellation Leo, Alpha Leonis—or Regulus, as it was named by Copernicus. The Romans referred to Regulus as Cor Leonis, the Lion's Heart. Regulus is a blue-white star with a companion star that is deep yellow in color and at magnitude 8.2 is 175 arc minutes from Regulus.

Alpha Librae, Zubenelgenubi. The brightest star in Libra is a nice binocular double of pale blue and yellow stars. The pale blue primary is magnitude 2.8 and the yellow secondary magnitude 5.2, with the two stars 231 arc seconds apart.

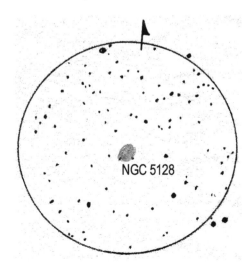

NGC 5128, Centaurus A
ILLUSTRATION BY DAVE BRANCHETT

Beta Librae. This interesting star, with what many observers see as a uniquely greenish tint, is also known as Zubeneschamali. Not everyone sees the greenish tint, but that is not the only bit of intrigue surrounding this star. The Greek astronomer Eratosthenes noted that it was so bright it even outshone Antares, which today is far brighter. A century later, the two stars were about the same brightness. So what happened? Did Zubeneschamali grow dimmer to its current magnitude of 2.6, or did Antares become brighter?

Sigma Librae. A magnitude 3.3 star that exhibits a dull ruddy orange color.

The Big Dipper. This asterism, part of the constellation Ursa Major, or the Big Bear, is fun and fascinating because of the seven stars that form it. They are all fairly bright, easy to find (as is the Big Dipper itself), and a great field to explore. Mizar and Alcor, the "second" star from the end of the handle of the Dipper, is actually two stars—a famous naked-eye optical double. It turns out that

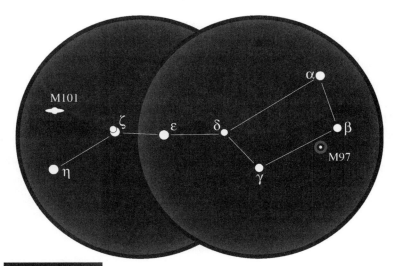

- ● Star
- ➤ Galaxy
- ◉ Globular Cluster
- ◉ Open Cluster
- ▬ Nebulosity
- ◉ Planetary Nebula

The Big Dipper, part of Ursa Major. Stars of the Big Dipper make fine introductory binocular objects. The diameter of one circle is approximately 23 degrees. ILLUSTRATION BY DAVID FRANTZ

Mizar is a true double, its magnitude 4 secondary a scant 14 arc seconds distant. Dubhe, the brightest star, is deep orange in color; Merak, Phecda, Megrez, Alioth, and Mizar are all a lovely blue-white; and Alkaid is a blue-silver. It turns out that five of the Big Dipper stars are moving through space together, part of an apparent group of about seventeen stars known to astronomers as Collinder 285. This cluster includes stars in Corona Borealis and Leo Minor.

M40. Also known as Winnecki 4, M40 is actually double stars of magnitude 9.0 and 9.3 separated by 50 arc seconds. M40 was listed as a nebula when first viewed in 1660 by Johannis Hevelius. Messier included the pair in his catalog in 1764. A. Winnecki listed it as a double star in his catalog in 1863, assigning it the number Winnecki 4. It is a tough object for 50-millimeter binoculars, yet a fun challenge.

VARIABLES

S Ant. An eclipsing binary star located in Antlia that varies from magnitude 6.4 to 6.9, with a 0.648-day period.

V CVn. A semiregular pulsating variable star found in Canes Venatici that varies from about magnitude 6.8 to 8.8, with a 196-day period. Its vivid red shade inspired the name La Superba.

R Hya. A Mira-type long-period variable star located in Hydra, varying from about magnitude 4.5 to 9.5, with a 388.87-day period.

NGC 3242/C59, the Ghost of Jupiter

PHOTO BY JERRY ARMSTRONG

SPRING DEEP-SKY OBJECTS
Select Objects Overview

CONSTELLATION	• Object: magnitude; type of object; size in degrees, arc minutes, or arc seconds as indicated *(As a reference, the moon and the sun are approximately 30 arc minutes across.)*
Bootes *the Herdsman*	• **NGC 5466:** 9.1 magnitude; globular cluster; 11 arc minutes in size
Canes Venatici *the Hunting Dogs*	• **M3:** 5.9 magnitude; globular cluster; about 16 arc minutes in diameter • **M51, the Whirlpool Galaxy:** 8.4 magnitude; spiral galaxy; 10 by 6 arc minutes in size • **M63:** 8.6 magnitude; spiral galaxy; 10 by 6 arc minutes • **M94:** 8.2 magnitude; spiral galaxy; 7 by 3 arc minutes • **M106:** 8.3 magnitude; spiral galaxy; 19 by 8 arc minutes in size • **Upgren 1:** About ten stars, the brightest of which are seventh magnitude; open cluster; 15 arc minutes in size
Coma Berenices *the Hair of Berenice*	• **M53:** 7.7 magnitude; globular cluster; 12.6 arc minutes in size • **M64:** 8.5 magnitude; spiral galaxy; 9.3 by 5.4 arc minutes in size • **Melotte 111, Coma Berenices Star Cluster:** About sixty stars, the brightest of which is fifth magnitude in this 5 degree open star cluster
Hydra *the Water Snake*	• **M48:** 5.8 magnitude; open star cluster; about 54 arc minutes in size • **M83, the Small Pinwheel:** 7.6 magnitude; spiral galaxy; 11 by 10 arc minutes in size • **NGC 3242/C59, Ghost of Jupiter:** 7.8 magnitude; planetary nebula; 0.3 arc minutes with the halo extending to 21 arc minutes
Leo *the Lion*	• **M65:** Challenging; 8.8 magnitude; spiral galaxy; 8 by 1.5 arc minutes in size • **M66:** Challenging; 9.0 magnitude; spiral galaxy; 8 by 2.5 arc minutes in size
Leo Minor *the Small Lion*	• **Sailboat:** Brightest star is seventh magnitude in this open cluster; 45 arc minutes in size

SPRING DEEP-SKY OBJECTS
Select Objects Overview *(continued)*

CONSTELLATION	• Object: magnitude; type of object; size in degrees, arc minutes, or arc seconds as indicated *(As a reference, the moon and the sun are approximately 30 arc minutes across.)*
Ursa Major *the Big Bear*	• **M81:** 6.9 magnitude; spiral galaxy; 21 by 10 arc minutes in size • **M82:** 8.4 magnitude; irregular galaxy; 9 by 4 arc minutes in size • **M97:** Challenging; 9.9 magnitude; planetary nebula; 3.4 by 3.3 arc minutes in size • **M101, Pinwheel Galaxy:** 7.9 magnitude; spiral galaxy; 22 arc minutes in diameter
Virgo *the Virgin*	• **M49:** 8.4 magnitude; elliptical galaxy; 9 by 7.5 arc minutes in size • M104, Sombrero Galaxy: 8.0 magnitude; spiral galaxy; 9 by 4 arc minutes in size

SPRING STARS AND DOUBLE STARS
Select Objects Overview

CONSTELLATION	• Object: type of object; magnitude(s); separation between the stars in arc seconds; and Position Angle in PA degrees [Year of Observation for Separation and PA data] for double/multiple stars
Bootes *the Herdsman*	• **Alpha Bootis, Arcturus:** –0.06 magnitude orange star. • **South 656:** Double stars of magnitude 6.9 and 7.4 separated by 86 arc seconds at PA 208° [2003] • **Iota Bootis:** Double stars of magnitude 4.9 and 7.5 separated by 40 arc seconds at PA 33° [1999] • **Mu Bootis:** Double stars of magnitude 4.3 and 6.5 separated by 107 arc seconds at PA 171° [1999]

SPRING STARS AND DOUBLE STARS
Select Objects Overview(*continued*)

CONSTELLATION	• Object: type of object; magnitude(s); separation between the stars in arc seconds; and Position Angle in PA degrees [Year of Observation for Separation and PA data] for double/multiple stars
Coma Berenices *the Hair of Berenice*	• **Alpha Comae Berenices:** Close binary star system of equal 5.1 magnitude stars • **17 Comae Berenices:** Double stars of magnitude 5.3 and 6.6 separated by 146 arc seconds at PA 250° [2002] • **32 and 33 Comae Berenices:** Double stars of magnitude 6.3 and 6.7 separated by 95 arc seconds at PA 49° [1922]
Hydra *the Water Snake*	• **27 Hydrae:** Double stars of magnitude 4.8 and 7.0 separated by 229 arc seconds at PA 212° [2002] • **Herschel 4465:** Double stars of magnitude 5.3 and 8.4 separated by 67 arc seconds at PA 44°[1919]
Leo *the Lion*	• **Alpha Leonis, Regulus, Regulus:** Double stars of magnitudes 1.4 and 8.2 separated by 175 arc seconds at PA 307° [1999]
Libra *the Scales*	• **Alpha Librae, Zubenelgenubi:** Double stars of blue 2.8 magnitude and yellow 5.2 magnitude separated by 231 arc seconds at PA 315° [2002] • **Beta Librae, Zubeneschamali:** Greenish-tinted 2.6 magnitude star • **Sigma Librae:** Dull ruddy orange color; 3.3 magnitude • **Iota Librae:** Double stars of magnitudes 4.5 and 9.8 separated by 58 arc seconds at PA 111° [2002]
Ursa Major *the Big Bear*	• **The Big Dipper:** Asterism composed of seven stars; Alkaid, Alioth, Dubhe, Megrez, Merak, Mizar, and Phecda • **Zeta and 80 Ursae Majoris, Mizar and Alcor:** Double stars of magnitudes 2.3 and 4.0 separated by 709 arc seconds at PA 152° [2002]

SPRING VARIABLE STARS
Select Objects Overview

CONSTELLATION	• Variable: type of variable; magnitude range; period (usually measured from one minima or primary eclipse to the next minima or primary eclipse); additional notes as appropriate
Antlia *the Air Pump*	• **S Ant:** Eclipsing binary 6.4–6.9 magnitude range with a 0.648-day period • **T Ant:** Cepheid with a 8.9–9.8 magnitude range with a 5.9-day period
Canes Venatici *the Hunting Dogs*	• **V CVn:** Semiregular pulsating around 6.8–8.8 magnitude range with a 196-day period
Hydra *the Water Snake*	• **R Hya:** Mira-type long-period around 4.5 to 9.5 magnitude with a 388.87 day • **U Hya:** Semiregular pulsating 7.0–9.4 magnitude range with a poorly established period
Leo the Lion	• **R Leo:** Mira-type long-period 5.8–10.0 magnitude range with a 309.95-day period
Leo Minor *the Small Lion*	• **R LMi:** Mira-type long-period 7.1–12.6 magnitude range with a 372.19-day period
Sextans *the Sextant*	• **RT Sex:** Semiregular pulsating 7.9–9.0 magnitude range with a poorly established period
Ursa Major *the Big Bear*	• **TX UMa:** Eclipsing binary 6.9–8.5 magnitude range with a 3.063-day period • **RY UMa:** Semiregular pulsating 6.7–8.3 magnitude range with a poorly established period
Virgo the Virgin	• **SS Vir:** Semiregular pulsating 6.8–8.9 magnitude range with a 364.14-day period • **R Vir:** Mira-type long-period 6.9–11.5 magnitude range with a 145.63-day period

9

Summer Objects

Summer brings warm temperatures and, for many stargazers, a yearning to observe. Some parts of the United States have their worst weather in the summer; clouds and thunderstorms can dominate for weeks on end. Yet when clear, the summer skies have a large number of terrific objects visible through even the smallest of binoculars.

The summer Milky Way and the stars and objects of Ophiuchus, Scorpius, and Sagittarius to the south and Aquila and Cygnus overhead are a bounty for deep-sky observers. If you enjoy looking at deep-sky objects, Sagittarius is the mother lode! I have always loved to take a pair of binoculars and sweep through these rich regions; I have spent hours searching out the jewels here.

Books have been written just on deep-sky objects, stars, double stars, eclipsing binaries, and variable stars. The highlights of the entire summer season will amaze the new skywatcher and never bore the seasoned observer.

The constellations of summer are many: Aquila, the Eagle; Corona Borealis, the Northern Crown; Cygnus, the Swan; Delphinus, the Dolphin; Draco, the Dragon; Equuleus, the Colt or Foal; Hercules, the Kneeler; Lyra, the Lyre; Ophiuchus, the Serpent-Bearer; Sagitta, the Arrow; Sagittarius, the Archer; Scorpius, the Scorpion; Scutum, the Shield; Serpens, the Snake; Ursa Minor, the Little Bear; and Vulpecula, the Little Fox. In our grouping, the summer zodiac includes Ophiuchus, Sagittarius, and Scorpius. You might be surprised to find Ophiuchus in the zodiac; the way

An all-sky photograph of the glorious summer Milky Way PHOTO BY CARTER ROBERTS

the sky has been configured with constellation boundaries is such that the zodiac crosses Ophiuchus.

Serpens is interesting because it is the only constellation divided in two by another constellation, Ophiuchus. You will often see references to Serpens Caput and Serpens Cauda. Serpens Caput is the head of the snake, Cauda is the tail. Together, though divided, they make up Serpens, the Snake.

The constellations low in the south include Ara, the Altar; Corona Australis, the Southern Crown; Lupus, the Wolf; Norma, the Carpenter's Square; and Telescopium, the Telescope.

DEEP-SKY OBJECTS

Aquila and the Milky Way. The constellation Aquila is not rich in identified deep-sky objects, so you might be tempted to ignore it. Take your binoculars and scan Aquila, however. You will find a simply wonderful view of the Milky Way. Also note an absence of stars, called the Great Rift.

Cygnus and the Milky Way. As with the Milky Way in Aquila, but even better, take your binoculars and scan Cygnus. You will find the best view of the Milky Way north of the Celestial Equator. You will also see the absence of stars known as Great Rift.

NGC 6826/C15, Blinking Nebula. A magnitude 8.8 planetary nebula in Cygnus that is 30 arc seconds in size. This nebula has a tendency to appear as though it is blinking on and off because of its size and brightness.

NGC 6543/C6, Cat's Eye Nebula. A magnitude 8.1 planetary nebula in Draco that is 18 arc seconds in size. Look for the lovely bluish tint.

Sagittarius and the Milky Way. You are in for a treat as you train your binoculars on this part of the sky. As a child seeing it for the first time, I wondered if it was going to rain because all these clouds were moving in! The region is also rich in deep-sky objects; a few are mentioned in the text, with many of the best objects, some challenging, listed in the Summer Objects table.

M13. The famous magnitude 5.3 globular cluster in Hercules, 16.6 arc minutes in diameter. Many note M13 as the best globular visible in the Northern Hemisphere.

M8, Lagoon Nebula. A magnitude 3 open cluster with nebulosity in Sagittarius, 60 by 35 arc minutes in size. It's considered to be one of the premier summer nebulae.

M22. A spectacular magnitude 5.2 globular cluster in Sagittarius, 24 arc minutes in diameter. M22 is about 50 percent larger than the acclaimed M13 globular in Hercules.

M23. A magnitude 5.5 open cluster in Sagittarius composed of around 120 stars, 27 arc minutes in size.

M6. A magnitude 4.2 open cluster in Scorpius, 15 arc minutes in size. This is a simply wonderful open cluster.

M7. A magnitude 2.8 open cluster also in Scorpius, 80 arc minutes in size. Even better than its close neighbor M6.

M16, Eagle Nebula and Cluster. A magnitude 6 open cluster with nebulosity in Serpens, 7 arc minutes in size. The stars that

Collinder 399, the Coat Hanger. This open cluster, in the center of the photo, takes its name from its hangerlike shape.
PHOTO BY VIC WINTER

A star-rich region in Sagittarius, including M8, the Lagoon Nebula, M20, the Trifid Nebula, and a "guest"—the bright planet Mars PHOTO BY VIC WINTER

make up M16 will be obvious; the nebulosity that forms the famous Eagle Nebula will be much more challenging in binoculars.

Collinder 399, the Coat Hanger. A magnitude 3.6 open cluster in Vulpecula, about 60 arc minutes in size. This is a nice and large open cluster that to my eyes does look somewhat like a coat hanger.

M27, Dumbbell Nebula. This is a very challenging magnitude 7.3 planetary nebula in Vulpecula, 8 by 5.7 arc minutes in size. The Dumbbell is one of the finest planetaries and worth a try, especially if you are using giant binoculars.

The bright star Deneb, the North America Nebula, along with the other splendors of Cygnus PHOTO BY VIC WINTER

STARS

Alpha, Beta, and Gamma Aquilae. About a 5-degree line of three bright stars in Aquila. Alpha, or Altair, is bluish; Beta an off-white; and Gamma a ruddy orange. The three stars were originally called Altair for "the whole group," a Bedouin term.

Alpha Cygni, Deneb. A bright magnitude 1.3 star in Cygnus. You will love the purity of Deneb's color.

Beta Cygni, Alberio. Gold and sapphire double stars of magnitudes 3.1 and 5.1 in Cygnus, separated by 34 arc seconds at PA 54 degrees.

Delphinus. This is one case where the constellation actually looks as advertised. It is wonderful in a pair of 7-power binoculars. Look for the brighter seven stars of around magnitude 4 to 5 in an area about 6 degrees in length.

Nu Draconis. Pure white double stars each of magnitude 4.9 in Draco, separated by 61 arc seconds at a position angle of 312 degrees.

Alpha Lyrae, Vega. A beautiful blue-white magnitude 0.0 star in Lyra. Vega, Altair in Aquila, and Deneb in Cygnus form the easy-to-recognize asterism known as the Summer Triangle.

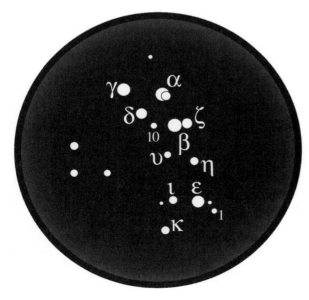

Stars in the constellation Delphinus. The circle is about 11 degrees in diameter.

ILLUSTRATION BY
DAVID FRANTZ

Sagittarius, occasionally known as "the teapot," is a rich summer constellation. The circle is about 35 degrees in diameter.

ILLUSTRATION BY DAVID FRANTZ

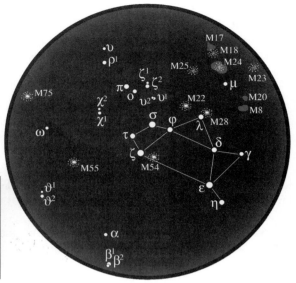

Delta Lyrae. Blue-white magnitude 5.6 and orange magnitude 4.5 double stars in Lyra, separated by 630 arc seconds

Epsilon Lyrae, the Double-Double. Four stars compose this famous double, thus the phrase "Double-Double." The first pair of stars is separated from the second pair of stars by 209 arc seconds at a position angle of 173 degrees. The first pair of stars has magnitudes of 4.7 and 5.8 and is itself separated by 2.44 arc seconds at a position angle of 349 degrees. The second pair has magnitudes of 5.3 and 5.5 and is separated by 2.32 arc seconds at a position angle of 83 degrees. With binoculars, you will be able to split the pairs separated by 209 arc seconds. But to split all four stars requires high magnification, i.e., a telescope. The Double-Double is often a favorite of observers showing off double star systems through their telescopes.

Sagittarius. This constellation is officially an archer, but most recognize it because of its asterism as a teapot. You can easily trace the spout, lid, handle, and the pot itself.

Alpha Scorpii, Antares. A brilliant ruddy orangish magnitude 1.0 star. Antares is from the Greek, meaning the opposite or rival of Ares, or Mars.

Theta Serpentis. Double stars of magnitudes 4.6 and 5.0 in Serpens Cauda (snake's tail), separated by 22 arc seconds at a position angle of 104 degrees. You will need binoculars of at least 10 power to split Theta Serpentis, a pleasing pair of blue stars that are similar in brightness.

The Little Dipper. An asterism made up of magnitude 2 to 5 stars; part of Ursa Minor, or the Little Bear. At the end of the dipper's handle is Alpha Ursa Minoris, or Polaris, the North Celestial Pole star. At magnitude 2.11 in Ursa Minor, this might be one of the best known but least recognized stars in the sky. Polaris comes from the Latin *stella polaris,* simply meaning "pole star." The two brightest bowl stars in the Little Dipper are the chrome-orange Beta Ursa Minoris, or Kochab, magnitude 2.1, and the blue-white Gamma Ursa Minoris, magnitude 3.0. These two stars are separated by about 3 degrees and are a great contrasting pair viewed with 7×50s.

VARIABLES

R CrB. A prototype of the R Corona Borealis type eruptive variable found in Corona Borealis which varies from magnitude 5.7 to 14.8. R CrB is its own namesake class of variable; variations in brightness are unpredictable and irregular.

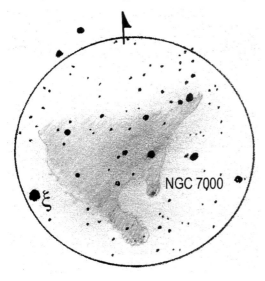

NGC 7000, the North America Nebula. It's best seen through 7×50s. ILLUSTRATION BY DAVE BRANCHETT

T CrB. Recurrent nova located in Corona Borealis that varies greatly from magnitude 2.0 to 10.8, with an irregular period.

X Cyg. A classical Cepheid located in Cygnus that varies from magnitude 5.9 to 6.9, with a 16.38-day period.

R Lyr. A semiregular pulsating variable found in Lyra that varies from magnitude 3.88 to 5.0, with a 46-day period.

RR Lyr. A RR Lyr–type pulsating variable in Lyra varying from magnitude 7.06 to 8.12, with a 0.5669-day period. RR Lyr is its own namesake class of variable stars that are pulsating white giants.

U Oph. An eclipsing binary found in Ophiuchus that varies from magnitude 5.8 to 6.6, with a 1.677-day period.

Y Sgr. Another classical Cepheid located in Sagittarius that varies from magnitude 5.4 to 6.1, with a 5.77-day period.

R Sct. A RV Tau–type pulsating variable in Scutum varying from magnitude 4.5 to 8.2, with a 146.5-day period.

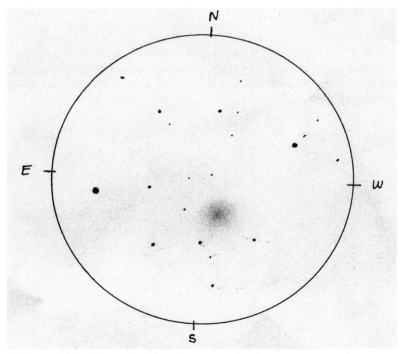

M4, a globular cluster in Scorpius ILLUSTRATION BY GONZALO VARGAS

SUMMER DEEP-SKY OBJECTS
Select Objects Overview

CONSTELLATION	• Object: magnitude; type of object; size in degrees, arc minutes, or arc seconds as indicated (*As a reference, the moon and the sun are approximately 30 arc minutes across.*)
Aquila *the Eagle*	• **Barnard 142 and 143, Fish on a Platter Nebula:** A challenging dark nebula 30 arc minutes in diameter with "fish tail" or horns
Cygnus *the Swan*	• **Barnard 168/IC 5146, Cocoon Nebula:** Dark nebula; 100 by 10 arc minutes in size • **M29:** 6.6 magnitude; open cluster; 7 arc minutes in size • **M39:** 4.6 magnitude; open cluster; 32 arc minutes in size • **NGC 6826/C15, Blinking Nebula:** 8.8 magnitude; planetary nebula; 30 arc seconds in size • **NGC 6871:** 5.2 magnitude; open cluster; 20 arc minutes in size • **NGC 6960/C34, West Veil Nebula:** Supernova remnant; 70 by 6 arc minutes in size • **NGC 6992/5/C33, East Veil Nebula:** Supernova remnant; 60 by 8 arc minutes in size • **NGC 7000/C20, North America Nebula:** Challenging; dark nebula; 120 by 100 arc minutes in size; best in 7×50s • **NGC 7039:** 7.6 magnitude; open cluster; 25 arc minutes in size
Delphinus *the Dolphin*	• **NGC 6934/C47:** 8.9 magnitude; globular cluster; 5.9 arc minutes in diameter
Draco *the Dragon*	• **NGC 4236/C3:** 9.7 magnitude; spiral galaxy; 21 by 7 arc minutes in size • **NGC 6543/C6, Cat's Eye Nebula:** 8.1 magnitude; planetary nebula; 18 arc seconds in size
Hercules *the Kneeler*	• **M13:** 5.3 magnitude; globular cluster; 16.6 arc minutes in diameter • **M92:** 6.5 magnitude; globular cluster; about 11 arc minutes in diameter
Hydra *the Snake*	• **M83, the Small Pinwheel:** 7.5 magnitude; spiral galaxy; 11 by 10 arc minutes in size

SUMMER DEEP-SKY OBJECTS
Select Objects Overview (continued)

CONSTELLATION	• Object: magnitude; type of object; size in degrees, arc minutes, or arc seconds as indicated (As a reference, the moon and the sun are approximately 30 arc minutes across.)
Lyra the Lyre	• **M56:** 8.8 magnitude; globular cluster; about 7 arc minutes in diameter
	• **M57, Ring Nebula:** Tough but well-known; 8.8 magnitude; planetary nebula; about 1.4 by 1 arc minutes in size
	• **Stephenson 1, Delta Lyrae Cluster:** 3.8 magnitude; open cluster; 25 arc minutes in size
Ophiuchus the Serpent-Bearer	• **Barnard 59, 65, 66, 67, 78, Pipe Nebula:** Dark nebula; extending over 7 degrees
	• **Barnard 72, Snake Nebula:** Dark nebula extending from the NW corner of the Pipe Nebula's bowl; 30 arc minutes in size
	• **IC 4665:** 4.2 magnitude; open cluster; 41 arc minutes in size
	• **Melotte 186:** Open cluster comprised of fourth to eighth magnitude stars extending about 2 degrees
	• **M9:** 7.8 magnitude; globular cluster; 9.3 arc minutes in diameter
	• **M10:** 6.6 magnitude; globular cluster; about 15 arc minutes in diameter
	• **M12:** 6.8 magnitude; globular cluster; 14.5 arc minutes in diameter
	• **M14:** 7.6 magnitude; globular cluster; 11.7 arc minutes in diameter
	• **M19:** 6.8 magnitude; globular cluster; 13.5 arc minutes in diameter
	• **M62:** 6.7 magnitude; globular cluster; about 14 arc minutes in diameter
	• **NGC 6633:** 4.6 magnitude; open cluster; 27 arc minutes in size
Sagittarius the Archer	• **M8, Lagoon Nebula:** 3.0 magnitude; open cluster with nebulosity and dark lanes; about 60 by 35 arc minutes in size
	• **M17, Omega Nebula:** 6.0 magnitude; bright nebulosity; about 47 by 36 arc minutes in size

SUMMER DEEP-SKY OBJECTS
Select Objects Overview (continued)

CONSTELLATION	• Object: magnitude; type of object; size in degrees, arc minutes, or arc seconds as indicated (As a reference, the moon and the sun are approximately 30 arc minutes across.)
Sagittarius utes *the Archer* (continued)	• **M18:** 6.9 magnitude; open cluster; 9 arc minin size • **M20, Trifid Nebula:** 6.3 magnitude; nebulosity with dark nebula; about 29 by 27 arc minutes in size • **M21:** 5.5 magnitude; open cluster; 13 arc minutes in diameter • **M22:** 5.2 magnitude; globular cluster; 24 arc minutes in diameter • **M23:** 5.5 magnitude; open cluster; 27 arc minutes in size • **M24:** 2.5 magnitude; star cloud/Milky Way patch; 90 arc minutes in size • **M25:** 4.6 magnitude; open cluster; 40 arc minutes in size • **M28:** 6.9 magnitude; globular cluster; about 11 arc minutes in diameter • **M54:** 7.2 magnitude; globular cluster; 9 arc minutes in diameter • **M55:** 6.3 magnitude; globular cluster; about 19 arc minutes in diameter • **M75:** Challenging; 8.6 magnitude; globular cluster; 6 arc minutes in diameter • **NGC 6822/C57, Barnard's Galaxy:** Challenging; 8.8 magnitude; irregular galaxy; 10 by 9 arc minutes in size
Scorpius *the Scorpion*	• **M4:** 5.4 magnitude; globular cluster; 26.3 arc minutes in diameter • **M6:** 4.2 magnitude; open cluster; 15 arc minutes in size • **M7:** 2.8 magnitude; open cluster; 80 arc minutes in size • **M80:** 7.3 magnitude; globular cluster; 8.9 arc minutes in diameter • **NGC 6302/C69, Bug Nebula:** Very challenging; 9.6 magnitude; planetary nebula; 0.8 arc minutes in size

SUMMER DEEP-SKY OBJECTS
Select Objects Overview (continued)

CONSTELLATION	• Object: magnitude; type of object; size in degrees, arc minutes, or arc seconds as indicated (As a reference, the moon and the sun are approximately 30 arc minutes across.)
Scorpius *the Scorpion* (continued)	• **NGC 6124/C75:** 5.8 magnitude; open cluster; 29 arc minutes in size • **NGC 6231/C76:** 2.6 magnitude; open cluster; 15 arc minutes in size
Scutum *the Shield*	• **Barnard 103:** Dark nebulosity; about 40 arc minutes in size • **M11, Wild Duck Cluster:** 5.3 magnitude; open cluster; 14 arc minutes in size; surrounded by dark nebula • **M26:** Challenging; 8.0 magnitude; open cluster; 15 arc minutes in size
Serpens *the Snake*	• **IC 4756:** Sixth magnitude; open cluster; 52 arc minutes in size • **M5:** 5.7 magnitude; globular cluster; 17.4 arc minutes in diameter • **M16, Eagle Nebula and Cluster:** 6.0 magnitude; open cluster with nebulosity; 7 arc minutes in size
Vulpecula *the Little Fox*	• **Collinder 399, the Coat Hanger:** 3.6 magnitude; open cluster; about 60 arc minutes in size • **M27, Dumbbell Nebula:** Challenging; 7.3 magnitude; planetary nebula; about 8 by 5.7 arc minutes in size • **NGC 6940:** Sixth magnitude; open cluster; about 20 arc minutes in size

SUMMER STARS AND DOUBLE STARS
Select Objects Overview

CONSTELLATION	• Object: type of object; magnitude(s); separation between the stars in arc seconds and Position Angle in PA degrees [Year of Observation for Separation and PA data] for double/multiple stars
Aquila *the Eagle*	• **Alpha, Beta, and Gamma Aquilae:** About a 5-degree line of three bright stars; Alpha or Altair is bluish, and Gamma a ruddy orange • **Struve 178:** White and yellow double stars of magnitudes 5.6 and 7.7 separated by 90 arc seconds at PA 268° [2001]
Corona Borealis *the Northern Crown*	• **Herschel V 38:** Double stars of magnitudes 6.4 and 9.8 separated by 32 arc seconds at PA 17° [1998]
Cygnus *the Swan*	• **Alpha Cygni, Deneb:** Magnitude 1.3 • **Beta Cygni, Alberio:** Double stars of magnitudes 3.1 and 5.1 separated by 34 arc seconds at PA 54° [2001] • **Omicron Cygni:** Triple stars of magnitudes 3.8, 7.0, and 5.0 separated by 105 and 334 arc seconds at PA 173° and 323° from the 3.8 magnitude primary [2001] • **Struve 182:** Double stars of magnitudes 7.5 and 8.6 separated by 73 arc seconds at PA 299° [2001] • **Struve 207:** Double stars of magnitudes 6.4 and 8.0 separated by 87 arc seconds at PA 64° [2003] • **61 Cygni:** Double stars of magnitudes 5.2 and 6.1 separated by 31 arc seconds at PA 150° [2002]
Delphinus *the Dolphin*	• **Delphinus:** The constellation's brighter seven stars of around fourth to fifth magnitude and fainter stars in an area about 7° in length
Draco *the Dragon*	• **Nu Draconis:** Double stars each of magnitude 4.9 separated by 61 arc seconds at PA 312° [2002] • **Psi Draconis:** Double stars of magnitudes 4.9 and 6.1 separated by 30 arc seconds at PA 150° [1958]

SUMMER STARS AND DOUBLE STARS
Select Objects Overview (continued)

CONSTELLATION	• Object: type of object; magnitude(s); separation between the stars in arc seconds and Position Angle in PA degrees [Year of Observation for Separation and PA data] for double/multiple stars
Draco the Dragon (continued)	• **16 and 17 Draconis:** Triple stars of magnitudes 5.4 (primary), 6.4, and 5.5 separated by 3.4 and 90 arc seconds at PA 108° and 194° [1958] *Note that the separation of the magnitude 6.4 star from the primary is too close to be seen in binoculars*
Equuleus the Foal	• **Gamma Equulei:** Double stars of magnitudes 4.7 and 6.1 separated by 338 arc seconds at PA 152° [2001]
Hercules the Kneeler	• **Alpha Herculis:** Orange star; magnitude 2.1
Hydra the Snake	• **27 Hydrae:** Double stars of magnitudes 4.8 and 7.0 separated by 229 arc seconds at PA 212° [2002] • **Herschel 4465:** Double stars of magnitudes 5.4 and 8.3 separated by 66 arc seconds at PA 44° [1999]
Lyra the Lyre	• **Alpha Lyrae, Vega:** Blue-white star; magnitude 0.0 • **Beta Lyrae:** Double stars of magnitudes 3.6 and 6.7 separated by 46 arc seconds at PA 149° [2002] *Note that the Beta Lyrae primary is a variable* • **Delta Lyrae:** Optical double stars of magnitudes 5.6 and 4.5 separated by about 630 arc seconds; found in the Delta Lyrae Cluster • **Epsilon Lyrae, the famous Double-Double:** Pair of stars of magnitudes 4.7 and 5.8 separated by 2.44 arc seconds at PA 349°; the second pair of magnitudes 5.3 and 5.5 separated by 2.32 arc seconds at PA 83°; the first pair of stars is separated from the second pair of stars by 209 arc seconds at a PA of 173° [2001]; the pairs of stars visible in binoculars but to split the four stars requires magnification (see note in text)

SUMMER STARS AND DOUBLE STARS
Select Objects Overview (continued)

CONSTELLATION	Object: type of object; magnitude(s); separation between the stars in arc seconds and Position Angle in PA degrees [Year of Observation for Separation and PA data] for double/multiple stars
Lyra *the Lyre* (continued)	• **Zeta Lyrae:** Double stars of magnitudes 4.4 and 5.7 separated by 44 arc seconds at PA 150° [2002]
Ophiuchus *the Serpent-Bearer*	• **Rho Ophiuchi:** Triple stars of magnitudes 5, 6.8, and 7.3 separated by 156 and 151 arc seconds at PA 253° and 360° [1991] • **36 Ophiuchi:** Orange triplets of magnitudes 5.1, 6.5, and 7.8 separated by 733 and 267 arc seconds at PA 74° and 337° [1991] • **South 694:** Double stars of magnitudes 6.7 and 7.3 separated by 79 arc seconds at PA 237° [2003]
Sagitta *the Arrow*	• **Epsilon Sagittae:** Double stars of magnitudes 5.7 and 8.3 separated by 87 arc seconds at PA 82° [2002] • **Theta Sagittae:** Double stars of magnitudes 6.6 and 7.5 separated by 90 arc seconds at PA 222° [2000]
Sagittarius *the Archer*	• **54 Sagittarii:** Double stars of magnitudes 5.4 and 7.7 separated by 46 arc seconds at PA 42° [2002]
Scorpius *the Scorpion*	• **Alpha Scorpii, Antares;** Ruddy-orange magnitude 1.0 • **Nu Scorpii:** Double stars of magnitudes 4.4 and 6.3 separated by 41 arc seconds at PA 336° [2002] • **Van den Bos 1833:** Double stars of magnitudes 5.5 and 6.6 separated by 57 arc seconds at 20°; found in NGC 6231 [1999]
Serpens *the Snake*	• **Nu Serpentis:** Double stars of magnitudes 4.3 and 9.2 separated by 46 arc seconds at PA 26° [2002] • **Theta Serpentis:** Double stars of magnitudes 4.6 and 5.0 separated by 22 arc seconds at PA 104° [2002]

SUMMER STARS AND DOUBLE STARS
Select Objects Overview (continued)

CONSTELLATION	• Object: type of object; magnitude(s); separation between the stars in arc seconds and Position Angle in PA degrees [Year of Observation for Separation and PA data] for double/multiple stars
Ursa Minor *the Little Bear*	• **Alpha Ursa Minoris, Polaris:** The North Celestial Pole star; magnitude 2.11 • **The Little Dipper:** Asterism made up of second to fifth magnitude stars • **Beta and Gamma Ursa Minoris, Beta or Kochab:** Magnitude 2.1 is described as chrome-orange and Gamma, magnitude 3.0 is blue-white; separated by about 3 degrees

SUMMER VARIABLE STARS
Select Objects Overview

CONSTELLATION	• Variable: type of variable; magnitude range; period (usually measured from one minima or primary eclipse to the next minima or primary eclipse); additional notes as appropriate
Aquila *the Eagle*	• **R Aql:** Mira-type long-period around 6.1–11.5 magnitude range with a 284.2-day period • **V Aql:** Semiregular pulsating 6.6–8.4 magnitude range with a poorly established period
Corona Borealis *the Northern Crown*	• **R CrB:** R CrB–type eruptive 5.7–14.8 magnitude range • **SW CrB:** Semiregular pulsating 7.8–8.5 magnitude range with a poorly established period • **T CrB:** Recurrent nova 2.0–10.8 magnitude range with an irregular period • **U CrB:** Eclipsing binary 7.0–8.4 magnitude range with a 3.452-day period • **V CrB:** Mira-type long-period around 7.5–11.0 magnitude range with a 357.63-day period

SUMMER VARIABLE STARS
Select Objects Overview (continued)

CONSTELLATION	• Variable: type of variable; magnitude range; period (usually measured from one minima or primary eclipse to the next minima or primary eclipse); additional notes as appropriate
Cygnus *the Swan*	• **AF Cyg:** Semiregular pulsating 7.4–9.4 magnitude range with a 92.5-day period with a poorly established period • **CH Cyg:** Symbiotic eruptive + semiregular pulsating 6.4–8.7 magnitude range with a 97-day period • **P Cyg:** S Dor–type eruptive 3–6 magnitude range with a 296.5-day period • **RS Cyg:** Semiregular pulsating around 7.2–9.0 magnitude range with a 417.39-day period • **U Cyg:** Mira-type long-period around 7.2–10.7 magnitude range with a 463.24-day period • **V460 Cyg:** Semiregular pulsating 5.6–7 magnitude range with a poorly established period • **V1339 Cyg:** Semiregular pulsating 5.9–7.1 magnitude range with a 35-day period (uncertain) • **W Cyg:** Semiregular pulsating 5.5–7.0 magnitude range with a 131.1-day period • **X Cyg:** Classical Cepheid 5.9–6.9 magnitude range with a 16.38-day period
Delphinus *the Dolphin*	• **U Del:** Semiregular pulsating 5.9–7.7 magnitude range with a poorly established period
Lyra *the Lyre*	• **R Lyr:** Semiregular pulsating 3.88–5.0 magnitude range with a 46-day period • **RR Lyr:** RR Lyr–type pulsating 7.06–8.12 magnitude range with a 0.5669-day period
Ophiuchus *the Serpent-Bearer*	• **BF Oph:** Classical Cepheid 6.9–7.7 magnitude range • **Chi Oph:** Gamma Cas–type eruptive irregular 4.2 magnitude range; rapid light changes • **U Oph:** Eclipsing binary 5.8–6.6 magnitude range with a 1.677-day period • **V533 Oph:** SR 7.2–8.5 magnitude range • **X Oph:** Mira-type long-period around 6.8–8.8 magnitude range with a 328.85-day period

SUMMER VARIABLE STARS
Select Objects Overview (*continued*)

CONSTELLATION	• Variable: type of variable; magnitude range; period (usually measured from one minima or primary eclipse to the next minima or primary eclipse); additional notes as appropriate
Sagitta *the Arrow*	• **S Sge:** Classical Cepheid 5.2–6.0 magnitude range • **U Sge:** Eclipsing binary 6.6–10.1 magnitude range with a 3.381-day period
Sagittarius *the Archer*	• **BB Sgr:** Classical Cepheid 6.6–7.3 magnitude range • **RS Sgr:** Eclipsing binary 6.0–7.0 magnitude range • **U Sgr:** Classical Cepheid 6.3–7.2 magnitude range • **XX Sgr:** Cepheid 8.4–9.3 magnitude range • **V350 Sgr:** Classical Cepheid 7.1–7. magnitude range • **V356 Sgr:** Eclipsing binary 6.8–7. magnitude range • **W Sgr:** Classical Cepheid 4.3–5.1 magnitude range • **X Sgr:** Classical Cepheid 4.2–4.9 magnitude range • **Y Sgr:** Classical Cepheid 5.4–6.1 magnitude range with a 5.77-day period • **YZ Sgr:** Classical Cepheid 7.0–7.8 magnitude range
Scorpius *the Scorpion*	• **BM Sco:** Semiregular pulsating 6.8–8.7 magnitude range; in M6 • **RV Sco:** Classical Cepheid 6.6–7.5 magnitude range • **RY Sco:** Classical Cepheid 7.52–8.44 magnitude range with a 20.3132-day period • **V636 Sco:** Classical Cepheid 6.4–6. magnitude range • **V861 Sco:** Eclipsing binary 6.1–6. magnitude range

SUMMER VARIABLE STARS
Select Objects Overview (continued)

CONSTELLATION	• Variable: type of variable; magnitude range; period (usually measured from one minima or primary eclipse to the next minima or primary eclipse); additional notes as appropriate
Scutum *the Shield*	• **R Sct:** RV Tau–type pulsating 4.5–8.2 magnitude range with a 146.5-day period • **RZ Sct:** Eclipsing binary 7.3–8.8 magnitude range
Ursa Minor *the Little Bear*	• **V UMi:** Semiregular pulsating 8.8–9.9 magnitude range with a 72.0-day period
Vulpecula *the Little Fox*	• **FI Vul:** Slow irregular 8.6–10.3 magnitude range • **RS Vul:** Eclipsing binary 6.8–7.8 magnitude range • **SV Vul:** Classical Cepheid 6.7–7.8 magnitude range • **U Vul:** Classical Cepheid 6.7–7.5 magnitude range • **Z Vul:** Eclipsing binary 7.3–8.9 magnitude range

10

Fall Objects

Fall is a good time for skywatching. The air is a little cooler and perhaps a little clearer, with fewer summer thunderstorms. The fall constellations are fun and sometimes challenging to find. A few of them—Andromeda, Pegasus, and Cetus—are characters in wonderful mythological tales. Others are more recent discoveries, but their names are still the products of vivid imagination, such as Fornax, Sculptor, and Triangulum.

The cast of fall characters includes Andromeda, the Chained Princess; Aquarius, the Water Carrier; Aries, the Ram; Capricornus, the Sea Goat; Cassiopeia, the Queen; Cetus, the Sea Monster; Fornax, the Furnace; Pegasus, the Flying Horse; Pisces, the Fishes; Sculptor, the Sculptor's Workshop; and Triangulum, the Triangle. The fall zodiac includes the constellations Aquarius, Aries, Capricornus, and Pisces. Low in the south, look for Grus, the Crane; Microscopium, the Microscope; Phoenix, the Phoenix; and Piscus Austrinus, the Southern Fish.

To me, the summer skies have always seemed very much "alive" with objects. Fall is a little quieter, with many of the late-summer objects of the Milky Way, Sagittarius, and the Summer Triangle region of Altair-Deneb-Vega still visible. You will still find some wonderful sights in the fall sky. The W shape of Cassiopeia is easy to recognize; it may appear as an M or a 3, depending on its location in the sky. The Great Square of Pegasus is also easy to recognize and will lead you to a truly wonderful object, the Great Andromeda Galaxy.

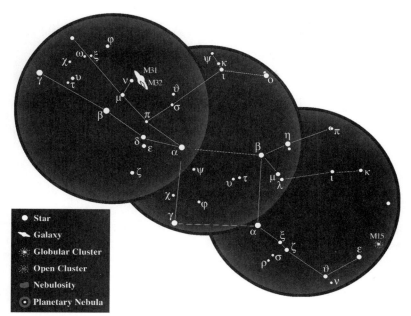

The prominent fall constellations Andromeda (left) and Pegasus (right). The square in Pegasus is easy to identify and allows you to pick out the rest of the constellation. Each circle is about 36 degrees in diameter.

ILLUSTRATION BY DAVID FRANTZ

DEEP-SKY OBJECTS

M31. The Great Andromeda Galaxy, a magnitude 3.4 spiral galaxy in the constellation of Andromeda that is 178 by 63 arc minutes in size. Great is simply an understatement; you can clearly see Andromeda with the naked eye, and binoculars will provide you with a breathtaking view. I enjoy the view of this spiral galaxy through my giant binoculars more than with a telescope, because I can see the entire galaxy in one field of view.

The Great Andromeda Galaxy is always one of the target objects for my college students, if the timing is just right. Many of the deep-sky objects are a challenge, and as viewed through binoculars or even a good-size amateur telescope, they do not look like their spectacular Hubble Telescope portraits. But M31 holds up to

M31, the Great Andromeda Galaxy
PHOTO BY VIC WINTER

the billing, always impressing the first-time viewer and reminding us how far away the Andromeda Galaxy is—2.4 million light years. The beauty of M31 just cannot be ignored.

Nearby are two challenging objects for binocular observers: M32 and M110. M32 is a magnitude 8.2 elliptical galaxy that is 8 by 6 arc minutes in size and a companion of M31. You will find M32 just south of and very close to the Great Andromeda Galaxy. M110, at magnitude 8 and 17 by 10 arc minutes in size, is north of the Great Andromeda Galaxy by about the width of M31.

Both of these ellipticals are challenging. When M32 is up, I often observe it through a pair of 7×50s to gauge the overall sky conditions. It's tough for me, but I can see it on a reasonably good night. Even though M110 is listed as lightly brighter than M32, I find it more of a challenge, often requiring higher magnification or larger binoculars. This has to do with the overall surface brightness of these two galaxies. Compare the sizes of M32 and M110: 8 by 6 arc minutes versus 17 by 10 arc minutes. Thus the light of M110 is more spread out than that of M32.

Sculptor Galaxy Group. There are several galaxies in the constellation Sculptor, including NCG 253/C65, the Sculptor Galaxy, NGC 300/C70, and NGC 55/C72. I have always found all of the Sculptor galaxies to be a bit challenging in 7×50s or even slightly larger binoculars, especially since they lie low in the southern skies (and I live in Florida). But they are worth a look, especially if you enjoy a challenge, live far enough south, and perhaps own a pair of giant binoculars.

STARS

Alpha Capricorni. Look in Capricornus for optical double stars of magnitudes 3.6 and 4.2, both of which appear yellowish. This optical double is separated by 378 arc seconds—a little over 6 arc minutes—at a position angle of 291 degrees.

Beta Capricorni. After examining Alpha Capricorni, continue your gazing in Capricornus to a pair of double stars of magnitudes 3.1 and 6.1, with the brighter star looking white and the dimmer a slight bluish color. You will find Beta Capricorni separated by 225 arc seconds at a position angle of 135 degrees.

Alpha Piscis Austrini, Fomalhaut. The brightest star in Piscis Austrinus, at magnitude 1.2, is joined as a double by a magnitude 6.5 star. Fomalhaut, seeming like an old southern friend, looks white, whereas its companion is reddish. The two stars are separated by 2 degrees, with the secondary due south of Fomalhaut.

VARIABLE

Omi Cet, Omicron Ceti, or Mira. A Mira-type long-period pulsating variable located in Cetus that varies from magnitude 3.4 to 9.3, with a 331.96-day period. Mira is the prototype for Mira-type variables and is a spectacular object to observe fading away as it becomes about 100 times fainter than its maximum brightness. This star actually disappears from eyesight!

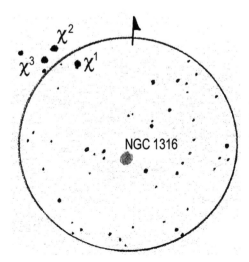

NGC 1316, Fornax A, an elliptical galaxy believed to have formed due to galactic cannibalism. ILLUSTRATION BY DAVE BRANCHETT

FALL DEEP-SKY OBJECTS
Select Objects Overview

CONSTELLATION	• Object: magnitude; type of object; size in degrees, arc minutes, or arc seconds as indicated *(As a reference, the moon and the sun are approximately 30 arc minutes across.)*
Andromeda *the Chained Princess*	• **M31, the Great Andromeda Galaxy:** 3.4 magnitude; spiral galaxy; 178 by 63 arc minutes in size • **M32:** 8.2 magnitude; elliptical galaxy; 8 by 6 arc minutes in size; companion of M31 • **M110:** 8.0 magnitude; elliptical galaxy; 17 by 10 arc minutes in size; a more challenging companion of M31 • **NGC 752/C28:** 5.7 magnitude; open cluster; 50 arc minutes in size • **NGC 7662/C22:** Challenging; 8.9 magnitude; planetary nebula; 32 by 28 arc seconds in size
Aquarius *the Water Carrier*	• **M2:** 6.3 magnitude; globular cluster; about 13 arc minutes in diameter • **NGC 7009/C55, Saturn Nebula:** Challenging; 7.6 magnitude; planetary nebula; 2.5 arc minutes in size • **NGC 7293/C63, Helix Nebula:** 7.3 magnitude; planetary nebula; 13 arc minutes in size
Capricornus *the Sea Goat*	• **M30:** 6.9 magnitude; globular cluster; 11 arc minutes in diameter
Fornax *the Furnace*or	• **NGC 1316, Fornax A:** 8.9 magnitude; elliptical spiral galaxy; 4 by 3 arc minutes in size
Pegasus *the Flying Horse*	• **M15:** 6.0 magnitude; globular cluster; 12.3 arc minutes in diameter
Sculptor *the Sculptor's Workshop*	• **NCG 253/C65, Sculptor Galaxy:** 7.1 magnitude; spiral galaxy; 22 by 6 arc minutes in size • **NGC 300/C70:** 8.7 magnitude; spiral galaxy; 23 by 13 arc minutes in size • **NGC 55/C72:** 7.9 magnitude; spiral galaxy; 32 by 6 arc minutes in size; located in the Sculptor Group • **Blanco 1, Zeta Sculptor Cluster:** 4.5 magnitude; open cluster; 90 arc minutes in size

FALL DEEP-SKY OBJECTS
Select Objects Overview *(continued)*

CONSTELLATION	• Object: magnitude; type of object; size in degrees, arc minutes, or arc seconds as indicated *(As a reference, the moon and the sun are approximately 30 arc minutes across.)*
Triangulum *the Triangle*	• **M33, Pinwheel:** 5.7 magnitude; spiral galaxy; 73 by 45 arc minutes in size • **Collinder 21:** Challenging; eighth magnitude; open cluster; 6 arc minutes in size; larger/higher-magnification binoculars reveal more stars

FALL STARS AND DOUBLE STARS
Select Objects Overview

CONSTELLATION	• Object: type of object; magnitude(s); separation between the stars in arc seconds and Position Angle in PA degrees [Year of Observation for Separation and PA data] for double/multiple stars
Andromeda *the Chained Princess*	• **56 Andromedae:** Double stars of magnitude 5.7 and 5.9 separated by 200 arc seconds at PA 299° [2001]. The 5.9 magnitude star is also a double with a star of a challenging magnitude 9.4 separated by 203 arc seconds at PA 258° [2001].
Aries *the Ram*	• **Lambda Arietis:** Two white double stars of magnitudes 4.8 and 7.7 separated by 37 arc seconds at PA 47° [2001] • **30 Arietis:** Yellow and pale lilac double stars of magnitudes 6.5 and 7.0 separated by 38 arc seconds at PA 274° [1997]
Capricornus *the Sea Goat*	• **Alpha Capricorni:** Yellowish optical double stars of magnitudes 3.6 and 4.2 separated by 378 arc seconds at PA 291° • **Beta Capricorni:** A complex multiple star system with primary white and slight bluish double stars of magnitudes 3.1 and 6.1 separated by 207 arc seconds at PA 267° [2001]. The third magnitude star is also has a component star of a challenging magnitude 9.0 separated by 225 arc seconds at PA 135° [2002].

FALL STARS AND DOUBLE STARS
Select Objects Overview *(continued)*

CONSTELLATION	• Object: type of object; magnitude(s); separation between the stars in arc seconds and Position Angle in PA degrees [Year of Observation for Separation and PA data] for double/multiple stars
Cetus *the Sea Monster*	• **37 Ceti:** Challenging double stars of magnitudes 5.2 and 7.9 separated by 49 arc seconds at PA 331° [2001]
Fornax *the Furnace*	• **Chi Fornacis:** A nice triangular asterism of several fifth–sixth magnitude stars; about nine stars which appear "diamondlike" in binoculars
Pegasus *the Flying Horse*	• **Epsilon Pegasi:** Double stars of magnitudes 2.5 and 8.7 separated by 144 arc seconds at PA 64° [1996] • **3 Pegasi:** Double stars of magnitudes 6.2 and 7.7 separated by 40 arc seconds at PA 349° [1995]
Pisces *the Fishes*	• **77 Piscium:** Double stars of magnitudes 6.4 and 7.3 separated by 32 arc seconds at PA 83° [2003]
Piscis Austrinus *the Southern Fish*	• **Alpha Piscis Austrini, Fomalhaut:** A whitish star of 1.2 magnitude with a reddish 6.5 magnitude companion, separated by 2 degrees at a PA of approximately 180°

FALL VARIABLE STARS
Select Objects Overview

CONSTELLATION	• Variable: type of variable; magnitude range; period (usually measured from one minima or primary eclipse to the next minima or primary eclipse); additional notes as appropriate
Aquarius *the Water Carrier*	• **V Aqr:** Semiregular pulsating giant 7.6–9.4 magnitude range with a 244-day period
Cetus *the Sea Monster*	• **Omi Cet, Mira:** Mira-type long-period pulsating around 3.4–9.3 magnitude range with a 331.96-day period • **T Cet:** Semiregular pulsating supergiant; 7.1–7.4 magnitude range
Fornax *the Furnace*	• **S For:** Uncertain type around 5.6–8 magnitude range and undetermined periodicity
Pegasus *the Flying Horse*	• **AG Peg:** Nova-type eruptive system 6.0–9.4 magnitude range; remains at maximum for 10+ years with a slow fade rate • **TW Peg:** Semiregular pulsating 7.5–8.4 magnitude range with a poorly established period
Pisces *the Fishes*	• **Z Psc:** Semiregular pulsating 8.8–10.1 magnitude range with a poorly established period • **RT Psc:** Semiregular pulsating 8.2–10.4 magnitude range with a poorly established period
Sculptor *the Sculptor's Workshop*	• **Y Scl:** Semiregular pulsating 8.7–10.3 magnitude range with a poorly established period
Triangulum *the Triangle*	• **W Tri:** Semiregular pulsating supergiant 8.5–9.7 magnitude range with a 108-day period

II

Winter Objects

Many occasional skywatchers, and even some seasoned amateurs, do not like winter observing because of the weather, but winter brings a celestial sky that should simply not be ignored. It features some of the brightest stars, most easily recognized constellations, memorable asterisms, wonderful clusters, and a nebula that includes the word "great" in its title for a reason. The winter skies present us with many beautiful, bright, and varying stellar colors to view. Eight of the twenty-five first-magnitude stars are winter constellation stars.

Winter is highlighted by Orion, the Hunter, the first constellation I learned, which is likely also the case for many others. Orion is supported by a cast of bright characters including Taurus, the Bull; Canis Major, the Big Dog; Canis Minor, the Little Dog (but with a bright star); Auriga, the Charioteer; Gemini, the Twins; and Perseus, the Hero. In addition, the winter skies include Camelopardalis, the Giraffe; Cancer, the Crab; Eridanus, the River; Lepus, the Hare; Lynx, the Lynx; and Monoceros, the Unicorn. Three of these are in the zodiac: Gemini, Taurus, and Cancer, which is a faint zodiacal constellation. Low in the south, if you are lucky, you will find Caelum, the Graving Tool; Carina, the Ship's Keel; Columba, the Dove; Horologium, the Clock; Pictor, the Painter's Easel; Puppis, the Stern; Pyxis, the Mariner's Compass; and Reticulum, the Reticle.

If the winter weather in your area is extreme, you should consider attending the Winter Star Party in the Florida Keys one year. There are many great reasons to attend this annual event, started

The winter Milky Way. Orion is to the right of center; the bright star below it is Sirius. PHOTO BY VIC WINTER

by Tippy D'Auria in 1983. It is in the Florida Keys, which usually have good warm weather, and it is far enough south—around latitude 24 degrees N—that you will get an opportunity to see some of those wonderful Southern Hemisphere objects. Good speakers are always scheduled, and you will have a chance to see all sorts of

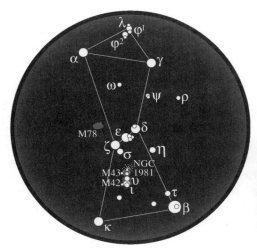

The constellation Orion is one of the easiest to recognize. The circle is approximately 22 degrees in diameter. ILLUSTRATION BY DAVID FRANTZ

● Star
◣ Galaxy
✴ Globular Cluster
✴ Open Cluster
▬ Nebulosity
◉ Planetary Nebula

astronomical equipment, including large binoculars and telescopes. Most attendees, which average around 500, welcome others to look through their optics. You need to register in advance and hope you can get in, because the Winter Star Party is popular.

If you are relegated to a frigid winter for now, fix a thermos of hot chocolate, bundle up, grab those binoculars, and find some clear skies. You have some observing to do!

DEEP-SKY OBJECTS

M36, M37, and M38. These three open clusters in Auriga always attract me, maybe because they are so close together in a nearly straight line, or perhaps because they are so different in appearance. M36 at magnitude 6.0 is the smallest yet the brightest, and I can resolve more of this open cluster. M37 looks like a hazy patch. M38, about the same size as M37, has a few stars that I can resolve.

M44. Often better known as Praesepe or the Beehive Cluster, this open cluster in Cancer is magnitude 3.1 and quite lovely in binoculars. It is large, at 95 arc minutes, and very easy to view with 7×35s or 7×50s. This is one that you will want to spend some time observing; it is quite pretty.

M67. This is another open cluster, also in Cancer, at magnitude 6.0. Even though it is not one of the brighter open clusters in the Northern Hemisphere, it is a nice and fairly easy 7×50 binocular object that is about 15 arc minutes in size.

NGC 2237 and 2244. The Rosette Nebula and Cluster consists of a magnitude 4.8 naked-eye open cluster in Monoceros that is 24 arc minutes in size surrounded by an 80 by 60 arc minute nebula. This is a really special view; you can see this nice open cluster surrounded by a dim glow.

M42 and M43. Located in the Sword of Orion, the Great Orion Nebula, or M42, is an easy naked-eye object, visible as a hazy patch. With a pair of 7×50s or 10×50s, you begin to see the photographic splendor of M42 and details are revealed. You might note a winglike appearance. The Great Orion Nebula is a magnitude 3.7 diffuse nebula 85 by 60 arc minutes in size. You can look at the Great Orion Nebula statistics, such as its brightness and size, or at photographs such as the one in chapter 7, taken by Vic Winter, but nothing prepares you for your first glimpse of this nebula. People's

This photograph focuses on Orion. M42 is easily visible, as is Barnard's Loop.
PHOTO BY VIC WINTER

reactions are similar to those they have when they first see the moon or Saturn through a telescope.

The Great Orion Nebula will amaze you night after night; the more you become familiar with M42 and the entire Sword of Orion region, the more you will see and enjoy. Within the region are a number of stars, some of which are discussed under Orion, below. M43 is a small, nebulous region separate from M42; many observe it thinking it is also a part of M42.

NGC 869 and 884/C14. The Double Cluster in Perseus is a magnitude 4.3 pair of open clusters that are each 30 arc minutes in size. How Charles Messier missed the Double Cluster is beyond me; they are so spectacular that even as you scan the sky without visual aid, you are drawn to these patches. Apparently Hipparchus noted them in 150 B.C. with the naked eye. A pair of 7×50s will provide you with a wonderful view of the Double Cluster; NGC 869, the western cluster in the pair, is made up of around 200 stars, and NGC 884, the eastern cluster, has about 150 stars. Spend some time with the Double Cluster, which will provide many evenings of pleasurable viewing and renewal with what will begin to feel like old friends.

C41. The Hyades is a magnitude 0.5 open cluster in Taurus and one of the closest to Earth. It is a terrific naked-eye and binocular object. This V-shaped object is 5.5 degrees or 330 arc minutes across. It is the face of Taurus the Bull and actually larger than the V-shaped asterism. Aldebaran, the bright golden-orange star, is

The Double Cluster in Perseus, a favorite binocular object
PHOTO BY VIC WINTER

one of the prominent winter stars. Aldebaran is not a physical part of the Hyades Association, as it is only about half as far away, yet it is associated with the cluster because of its location.

M45. The Pleiades is a magnitude 1.5 open cluster in Taurus that is 110 arc minutes in size. It is a wonderful binocular object, whether you are using 6×40s or 11×80s. Many people mistake it for the Little Dipper because of its shape. You have seen it before—on the back of Subaru automobiles. There is much to explore in the Pleiades. First note the similarities in stellar brightness and bluish silver color. The members of the Pleiades are relatively young stars. With giant binoculars, you can usually glimpse some nebulosity surrounding Merope. There are also some fine double stars in the Pleiades.

STARS

Alpha Aurigae, Capella. At magnitude 0.1, this is the sixth-brightest star in the heavens and has a beautiful yellow color. It is fun to compare Capella through binoculars to other stars in Auriga. For example, you will find a group referred to as "the Kids" roughly southwest of Capella. These three stars forming a small triangle are Epsilon, Zeta, and Eta Aurigae. The Greeks and Romans only noted two of the stars, Zeta and Eta, which are about 0.75 degrees apart. Eta is a magnitude 3.2 bluish silver star, and Zeta a magnitude 3.8 orange star that is actually an eclipsing binary.

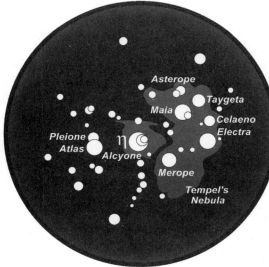

M45, the Pleiades. Note the nebulosity within the open cluster. PHOTO BY VIC WINTER; MAP BY DAVID FRANTZ

Canis Major. At magnitude −1.46, Alpha Canis Majoris, or Sirius, is a brilliant blue-white star and the second-brightest star (second only to the sun) in our sky. Sirius truly looks like a sparkling diamond through a pair of binoculars. Check out Sirius when it is close to the horizon; you will see a dazzling light show due to Earth's atmosphere. Other stars to observe in Canis Major include

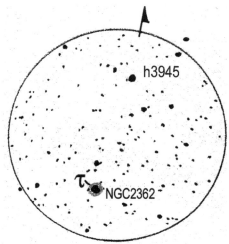

NGC 2362/C64, the Mexican
Jumping Star (Tau Canis
Majoris) ILLUSTRATION BY DAVE
BRANCHETT

Delta Canis Majoris, a yellow magnitude 1.8 star; Eta, a bluish silver magnitude 2.4 star; Sigma, a chrome red-orange magnitude 3.5 star; and Omicron 1, a chrome orange magnitude 3.8 star.

Alpha Canis Minoris, Procyon. This bright whitish magnitude 0.4 star literally leads the way for Sirius. In fact, Procyon is Greek for "before the dog." Check out Procyon's color, which some describe as an off-white or cream.

Cassiopeia "W" Asterism. The distinctive W or M shape of Cassiopeia makes it easy to find. You can explore the many open clusters found within Cassiopeia, one of the richest regions in our heavens. Binocular inspection will also reveal the Milky Way, often ignored yet very attractive in binoculars.

Alpha Carinae, Canopus. The third-brightest star in the sky (after the sun and Sirius) is a bright white magnitude –0.72 star in Carina, low in the south.

Alpha and Beta Geminorum, Castor and Pollux. Their brightness and color make these magnitude 1.57 and 1.14 stars in Gemini excellent binocular objects. Compare bluish white Castor to orange-yellow Pollux.

Gamma Leporis. A nice double star at magnitudes 3.6 and 6.3 that is separated by 97 arc seconds at a position angle of 350 degrees. The magnitude 3.7 star is yellow, and the magnitude 6.3

4

star appears to be somewhat orange. Gamma Leporis is easily split in a pair of 7×50s.

Alpha and 38 Lyncis. These make a nice comparison of magnitude 3.2 reddish Alpha and magnitude 3.8 bluish white 38 Lyncis about 2 degrees north of Alpha Lyncis.

Orion. Simply a treasure chest of binocular objects, Orion is probably the easiest constellation to recognize. Betelgeuse at magnitude 0.08 and Rigel at magnitude 0.12, on opposite sides of Orion, offer wonderful contrasting bright reddish and bluish stars. Also note other bright stars in Orion, such as Bellatrix, the right-shoulder star opposite Betelgeuse, and Saiph, the left-knee star opposite Rigel.

Orion's Belt Stars, known as Collinder 70, are a magnitude 0.4 open cluster 150 arc minutes in size. Delta Orionis, Mintaka, the faintest belt star, is actually a beautiful blue-white double star of magnitudes 2.4 and 6.8 separated by 53 arc seconds. Delta Orionis along with the other two bright belt stars, Epsilon and Zeta Orionis, Alnilam and Alnitak, make for a splendid trio of objects within a rich field of sixth- and seventh-magnitude stars. A telescope simply does not reveal the beauty of Orion's Belt like a pair of binoculars does. This area of the sky is as enjoyable as the Pleiades. By the way, the names Mintaka, Alnilam, and Alnitak are Arabic for "girdle," "string of pearls," and the "belt."

The Sword of Orion offers one of the finest deep-sky objects in the heavens—the Great Orion Nebula, M42 and M43—as well as a variety of other objects, such as Theta[1] and Theta[2] Orionis, double stars that both have magnitudes of 5.0 and are separated by 135 arc seconds at a position angle of 314 degrees. Theta[1] is the Trapezium, an object you must view sometime through a telescope.

Alpha Tauri, Aldebaran. This magnitude 0.8 star is worth inspecting through your binoculars. You will find a nice golden-orange star that leads the way to your easy inspection of the Hyades. Even though Aldebaran appears to be a part of the V-shaped Hyades, it is actually a foreground star and not physically associated with this cluster.

Pleiades. Beyond the beauty of the Pleiades, a number of double stars are visible here. Atlas and Pleione (BU Tau) are double stars at magnitudes 3.7 and 4.8 to 5.5 separated by 5 arc minutes at a position angle of 180 degrees. Pleione, BU Tau, is also a Gamma Cas–type variable star. Asterope is actually a double star at magnitudes 5.6 and 6.4 separated by about three minutes.

Theta¹ and Theta² Tauri. Double stars of magnitudes 3.4 and 3.8 separated by 337 arc seconds at PA 348 degrees.

Beta Tauri and Iota Auriga. These two stars are across the constellation border from each other, about 6 degrees apart. Beta Tauri, in Taurus, is a bluish silver magnitude 1.7 star; Iota Aurigae, in Auriga, is an orange magnitude 2.7 star. The pair makes for a fine color contrast in the field of view of a pair of 7 × 50s.

VARIABLES

Gamma Cas. A prototype of the Gamma Cas–type eruptive irregular variable found in Cassiopeia that goes from magnitude

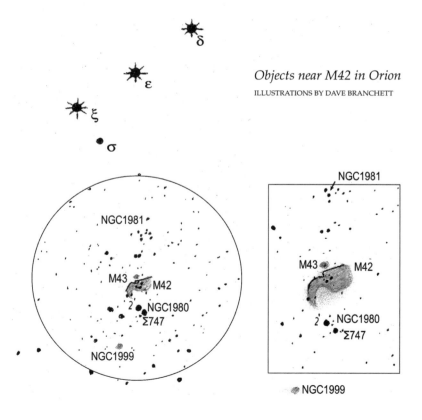

Objects near M42 in Orion
ILLUSTRATIONS BY DAVE BRANCHETT

1.6 to 3.0, with a 95-day period, which itself is considered to be uncertain.

Eta Gem. An irregular and eclipsing binary star in Gemini that varies from magnitude 3.15 to 3.9.

Alpha Ori. Better known as Betelgeuse, the brilliant orange star in Orion, this semiregular pulsating supergiant varies from magnitude 0.0 to 1.3, with a 335-day period.

Beta Per, Beta Persei, or Algol. An Algol-type eclipsing binary star in Perseus that goes from magnitude 2.1 to 3.4, with a 2.867-day period. Algol is the prototype for its class of close binary star systems. Algol is fun to watch simply because it's predictable and you can easily follow the change in brightness over a short period of time. This is one of my all-time favorite objects, and I enjoy showing Algol to the general public and my students, because they are amazed to see the variation in brightness and love the story and history as well.

Rho Per, Rho Persei. A semiregular pulsating variable in Perseus that varies from magnitude 3.3 to 4.0. The challenge with Rho Persei and this class of variables is that the period is poorly established.

BU Tau. A Gamma Cas–type eruptive irregular variable in Taurus that varies from magnitude 4.8 to 5.5. Being an irregular means the period is not well established, so it's fun to check on a regular basis.

WINTER DEEP-SKY OBJECTS
Select Objects Overview

CONSTELLATION	• Object: magnitude; type of object; size in degrees, arc minutes, or arc seconds as indicated (*As a reference, the moon and the sun are approximately 30 arc minutes across.*)
Auriga *the Charioteer*	• **Collinder 62:** 4.2 magnitude; open cluster; 28 arc minutes in size • **M36:** 6.0 magnitude; open cluster; 12 arc minutes in size • **M37:** 5.6 magnitude; open cluster; 24 arc minutes in size • **M38:** seventh magnitude; open cluster; 21 arc minutes in size
Camelopardalis *the Giraffe*	• **NGC 1502:** 5.7 magnitude; open cluster; 8 arc minutes in size • **NGC 2403/C7:** 8.4 magnitude; spiral galaxy; 18 by 10 arc minutes in size • **Stock 23:** seventh–ninth magnitude; open cluster; 15 arc minutes in size
Cancer *the Crab*	• **M44:** Praesepe or the Beehive Cluster; 3.1 magnitude; open cluster; 95 arc minutes in size • **M67:** 6.0 magnitude; open cluster; 15 arc minutes in size
Canis Major *the Large Dog*	• **Collinder 121:** By Omicron Canis Majoris; 2.6 magnitude; open cluster; 50 arc minutes in size • **Collinder 140:** Fourth magnitude; open cluster; 42 arc minutes in size • **M41:** 4.5 magnitude; open cluster; 38 arc minutes in size • **NGC 2354:** 6.5 magnitude; open cluster; 20 arc minutes in size • **NGC 2360/C58:** 7.2 magnitude; open cluster; 13 arc minutes in size • **NGC 2362/C64, Mexican Jumping Star (Tau Canis Majoris):** 4.1 magnitude; open cluster; 8 arc minutes in size
Cassiopeia *the Queen*	• **Collinder 463:** 5.7 magnitude; open cluster; 36 arc minutes in size • **M52:** 6.9 magnitude; open cluster; 13 arc minutes in size

WINTER DEEP-SKY OBJECTS
Select Objects Overview (continued)

CONSTELLATION	• Object: magnitude; type of object; size in degrees, arc minutes, or arc seconds as indicated (As a reference, the moon and the sun are approximately 30 arc minutes across.)
Cassiopeia *the Queen* (continued)	• **M103:** 7.4 magnitude; open cluster; 6 arc minutes in size • **NGC 129:** 6.5 magnitude; open cluster; 21 arc minutes in size • **NGC 225:** Seventh magnitude; open cluster; 12 arc minutes in size • **NGC 457/C13:** 6.4 magnitude; open cluster; 13 arc minutes in size • **NGC 581:** 7.4 magnitude; open cluster; 6 arc minutes in size • **NGC 663/C10:** 7.1 magnitude; open cluster; 16 arc minutes in size • **NGC 1027:** 6.7 magnitude; open cluster; 20 arc minutes in size • **Stock 2:** 4.4 magnitude; open cluster; 60 arc minutes in size
Gemini *the Twins*	• **Collinder 89:** 5.7 magnitude; open cluster; 35 arc minutes in size • **M35:** 5.1 magnitude; open cluster; 28 arc minutes in size • **NGC 2392, Eskimo Nebula:** 8.3 magnitude; planetary nebula; 13 arc seconds in size
Lynx *the Lynx*	• **NGC 2683:** Challenging; 9.7 magnitude; spiral galaxy; 8 by 1 arc minutes in size
Monoceros *the Unicorn*	• **Collinder 91:** 6.4 magnitude; open cluster; 17 arc minutes in size • **Collinder 96:** 7.3 magnitude; open cluster; 8 arc minutes in size • **Collinder 97:** 5.4 magnitude; open cluster; 21 arc minutes in size • **Collinder 106:** 4.6 magnitude; open cluster; 45 arc minutes in size • **Collinder 107:** 5.1 magnitude; open cluster; 35 arc minutes in size • **M50:** 5.9 magnitude; open cluster; 16 arc minutes in size

WINTER DEEP-SKY OBJECTS
Select Objects Overview *(continued)*

CONSTELLATION	• Object: magnitude; type of object; size in degrees, arc minutes, or arc seconds as indicated *(As a reference, the moon and the sun are approximately 30 arc minutes across.)*
Monoceros *the Unicorn* (continued)	• **NGC 2232:** 3.9 magnitude; open cluster; 30 arc minutes in size • **NGC 2237-9 and 2244/C49 and C50, Rosette Nebula and Cluster:** 4.8 magnitude; open cluster; 24 arc minutes in size surrounded by an 80-by-60 arc minute nebula • **NGC 2264, Christmas Tree Cluster:** 3.9 magnitude; open cluster; 20 arc minutes in size • **NGC 2301:** 5.8 magnitude; open cluster; 12 arc minutes in size
Orion *the Hunter*	• **Collinder 65:** Third magnitude; open cluster; 220 arc minutes in size • **Collinder 69, Lambda Orionis Cluster:** 2.8 magnitude; open cluster; 65 arc minutes in size • **Collinder 70, Orion's Belt Stars:** 0.4 magnitude; open cluster; 150 arc minutes in size • **M42, Great Orion Nebula:** 3.7 magnitude; diffuse nebula; 85 by 60 arc minutes in size • **M43:** 6.8 magnitude; diffuse nebula; 20 by 15 arc minutes in size • **M78:** 8.0 magnitude; diffuse nebula; 8 by 6 arc minutes in size • **NGC 1981:** 4.6 magnitude; open cluster; 25 arc minutes in size • **NGC 1977:** 4.6 magnitude; diffuse nebula; 20 by 10 arc minutes in size • **NGC 2169:** 5.9 magnitude; open cluster; 7 arc minutes in size
Perseus *the Hero*	• **M34:** 5.2 magnitude; open cluster; 35 arc minutes in size • **NGC 869 and 884/C14, the Double Cluster in Perseus:** 4.3 magnitude; open clusters; each 30 arc minutes in size • **NGC 1499, California Nebula:** A sometimes quite challenging dark nebula 145 by 40 arc minutes in size

WINTER DEEP-SKY OBJECTS
Select Objects Overview (continued)

CONSTELLATION	• Object: magnitude; type of object; size in degrees, arc minutes, or arc seconds as indicated (As a reference, the moon and the sun are approximately 30 arc minutes across.)
Perseus the Hero (continued)	• **NGC 1528:** 6.4 magnitude; open cluster; 25 arc minutes in size • **NGC 1545:** 6.2 magnitude; open cluster; 18 arc minutes in size • **Trumpler 2:** 5.9 magnitude; open cluster; 20 arc minutes in size
Taurus the Bull	• **C41, Hyades:** 0.5 magnitude; open cluster; 330 arc minutes in size • **M1, Crab Nebula:** A challenging 8.0 magnitude supernova remnant 6 by 4 arc minutes in size • **M45, Pleiades:** 1.5 magnitude; open cluster; 110 arc minutes in size • **NGC 1647:** 6.3 magnitude; open cluster; 45 arc minutes in size • **NGC 1746:** Sixth magnitude; open cluster; 45 arc minutes in size • **NGC 1807:** Seventh magnitude; open cluster; 17 arc minutes in size • **NGC 1817:** 7.7 magnitude; open cluster; 15 arc minutes in size

WINTER STARS AND DOUBLE STARS
Select Objects Overview

CONSTELLATION	• Object: type of object; magnitude(s); separation between the stars in arc seconds and Position Angle in PA degrees [Year of Observation for Separation and PA data] for double/multiple stars
Carina *the Ship's Keel*	• **Alpha Carinae, Canopus:** A bright, white star of −0.72 magnitude; a challenge due to its far southern location
Auriga *the Charioteer*	• **Alpha Aurigae, Capella:** 0.1 magnitude • **Epsilon, Zeta and Eta Aurigae, the Kids:** Triangle-shaped asterism • **Iota Aurigae and Beta Tauri:** Iota Aurigae, an orange 2.7 magnitude star, is about six degrees from bluish silver 1.7 magnitude Beta Tauri
Camelopardalis *the Giraffe*	• **11 Camelopardalis:** Double stars of magnitudes 5.2 and 6.1 separated by 179 arc seconds at PA 9° [2002] • **Struve 36:** Double stars of magnitudes 6.9 and 8.3 separated by 46 arc seconds at PA 71° [1999] • **Struve 90:** Double stars of magnitudes 6.2 and 7.5 separated by 51 arc seconds at PA 81° [1999] • **South 436:** Double stars of magnitudes 6.5 and 7.2 separated by 57 arc seconds at PA 76° [2002]
Cancer *the Crab*	• **Burnham 584:** Stars within M44 of magnitudes 6.9 and 7.2 separated by 45 arc seconds at PA 157° [2002]. A third star in the system at about the same magnitude is 93 arc seconds from the magnitude 6.9 star forming a nice triangle visible in 7× binoculars • **Iota Cancri:** Double stars of magnitudes 4.1 and 6.0 separated by 31 arc seconds at 307° [2002]
Canis Major *the Large Dog*	• **Alpha Canis Majoris, Sirius:** Magnitude −1.46; a brilliant blue-white star • **Delta Canis Majoris:** 1.8 magnitude star • **Eta Canis Majoris:** 2.4 magnitude star • **Sigma Canis Majoris:** 3.5 magnitude star • **Omicron 1 Canis Majoris:** 3.8 magnitude star • **Dunlop 47:** Double stars of magnitudes 5.4 and 7.6 separated by 99 arc seconds at PA 343° [1999]

WINTER STARS AND DOUBLE STARS
Select Objects Overview (continued)

CONSTELLATION	• Object: type of object; magnitude(s); separation between the stars in arc seconds and Position Angle in PA degrees [Year of Observation for Separation and PA data] for double/multiple stars
Canis Minor *the Small Dog*	• **Procyon:** Bright whitish star of magnitude 0.4
Cassiopeia *the Queen*	• **"W" Asterism:** The distinctive W or M shape of Cassiopeia • **Struve 26:** Double stars of magnitudes 6.9 and 7.2 separated by 63 arc seconds at PA 202° [2002]
Eridanus *the River*	• **Herschel 3628:** Double stars of magnitudes 7.2 and 8.0 separated by 50 arc seconds at PA 50° [2000] • **Omicron Eridani:** Double stars of magnitudes 4.4 and 9.7 separated by 83 arc seconds at PA 104° [2002]; the secondary is a challenge
Gemini *the Twins*	• **Alpha Geminorum, Castor:** 1.57 magnitude star • **Beta Geminorum, Pollux:** 1.14 magnitude star • **Nu Geminorum:** Double stars of magnitudes 4.1 and 8.0 separated by 111 arc seconds at PA 331° [2002]
Lepus *the Hare*	• **Gamma Leporis:** Double stars of magnitudes 3.6 and 6.3 separated by 97 arc seconds at PA 350° [1999]
Lynx *the Lynx*	• **Alpha and 38 Lyncis:** Third magnitude stars about 2 degrees apart • **5 Lyncis:** Double stars of magnitudes 5.2 and 7.9 separated by 96 arc seconds at PA 272° [2002]
Monoceros *the Unicorn*	• **Zeta Monocerotis:** Double stars of magnitudes 4.4 and 8.8 separated by 65 arc seconds at PA 247° [2002]
Orion *the Hunter*	• **Delta Orionis, Mintaka:** Double stars of magnitudes 2.4 and 6.8 in the belt of Orion separated by 53 arc seconds at PA 0° [2003]

WINTER STARS AND DOUBLE STARS
Select Objects Overview *(continued)*

CONSTELLATION	• Object: type of object; magnitude(s); separation between the stars in arc seconds and Position Angle in PA degrees [Year of Observation for Separation and PA data] for double/multiple stars
Orion *the Hunter* (continued)	• **Struve 747:** Double stars of magnitudes 4.7 and 5.5 separated by 36 arc seconds at PA 224° [2003]
Orion *the Hunter* (continued)	• **Theta¹ and Theta² Orionis:** Multiple stars. Theta¹ and Theta² Orionis are magnitudes 5.0 and 5.0 separated by 135 arc seconds at PA 314° [2000]. Theta² Orionis is magnitudes 5.0 and 6.2 separated by 53 arc seconds at PA 93° [2002]. Theta¹ is the Trapezium.
Perseus *the Hero*	• **57 Persei:** Double stars of magnitudes 6.1 and 6.8 separated by 120 arc seconds at PA 198° [2003]
Taurus *the Bull*	• **Eta Tauri, Alcyone:** Quadruple stars of magnitudes 2.8, 6.3, 8.3, and 8.7 separated by 118, 181, and 191 arc seconds at PA 290°, 314°, and 298° [2002–2003] • **Alpha Tauri, Aldebaran:** First magnitude star • **27 and BU Tauri, Atlas and Pleione:** Double stars of magnitudes 3.7 and 4.8–5.5 separated by 5 arc minutes at PA 180° • **Asterope:** Double stars of magnitudes 5.6 and 6.4 separated by 168 arc seconds • **Kappa Tauri:** Double stars of magnitudes 4.2 and 5.3 separated by 340 arc seconds at PA 174° [2002] • **Theta¹ and Theta² Tauri:** Double stars of magnitudes 3.4 and 3.8 separated by 337 arc seconds at PA 348° [2002] • **88 Tauri:** Double stars of magnitudes 4.3 and 7.8 separated by 69 arc seconds at PA 300° [2003] • **Struve 67:** Double stars of magnitudes 6.0 and 8.3 separated by 75 arc seconds at PA 162° [2002]

WINTER VARIABLE STARS
Select Objects Overview

CONSTELLATION	• Variable: type of variable; magnitude range; period (usually measured from one minima or primary eclipse to the next minima or primary eclipse); additional notes as appropriate
Auriga *the Charioteer*	• **AB Aur:** Irregular 6.9–8.4 magnitude range • **RT Aur:** Delta Cepheid–type 5.0–5.8 magnitude range with a 3.72-day period • **UU Aur:** Semiregular pulsating 8.2–10.0 magnitude range with a 234-day period
Cancer *the Crab*	• **R Cnc:** Mira-type long-period around 6.8–11.2 magnitude range with a 361.6-day period
Canis Major *the Large Dog*	• **R CMa:** Algol-type eclipsing binary 5.7–6.3 magnitude range with a 1.136-day period; close binary system
Cassiopeia *the Queen*	• **Gamma Cas:** Gamma Cas–type eruptive irregular 1.6–3.0 magnitude range with a 95-day period (uncertain) • **Rho Cas:** Semiregular pulsating 4.1–6.2 magnitude range • **RZ Cas:** Algol-type eclipsing binary 6.4–7.8 magnitude range with a 1.195-day period; close binary system • **TV Cas:** Algol-type eclipsing binary 7.3–8.4 magnitude range with a 1.813-day period; close binary system
Eridanus *the River*	• **RR Eri:** Semiregular pulsating 6.8–8.2 magnitude range with a 97-day period • **Z Eri:** Semiregular pulsating 6.3–7.9 magnitude range with a 80-day period
Gemini *the Twins*	• **BU Gem:** Irregular supergiant 5.7–8.1 magnitude range • **Eta Gem:** Irregular and eclipsing binary 3.15–3.9 magnitude range
Lepus *the Hare*	• **S Lep:** Semiregular pulsating 6.0–7.6 magnitude range with a poorly established period
Lynx *the Lynx*	• **SV Lyn:** Semiregular pulsating 8.2–9.1 magnitude range with a poorly established period

WINTER VARIABLE STARS
Select Objects Overview (continued)

CONSTELLATION	• Variable: type of variable; magnitude range; period (usually measured from one minima or primary eclipse to the next minima or primary eclipse); additional notes as appropriate
Monoceros *the Unicorn*	• **U Mon:** RV Tau–type pulsating 5.5–7.7 magnitude range with a 91.32-day period
Orion *the Hunter*	• **Alpha Ori, Betelgeuse:** Semiregular pulsating supergiant 0.0–1.3 magnitude range with a 335-day period • **KX Ori:** Irregular 6.9–8.1 magnitude range; poorly studied with unknown features
Perseus *the Hero*	• **Beta Per, Algol:** Algol-type eclipsing binary 2.1–3.4 magnitude range with a 2.867-day period; close binary system • **Rho Per:** Semiregular pulsating 3.3–4.0 magnitude range with a poorly established period • **KK Per:** Irregular 6.6–7.9 magnitude range with a poorly defined or insufficiently studied period • **IZ Per:** Eclipsing binary 7.8–9.0 magnitude range
Taurus *the Bull*	• **BU Tau:** Gamma Cas–type eruptive irregular 4.8–5.5 magnitude range • **CD Tau:** Algol-type eclipsing binary 6.8–7.3 magnitude range • **HU Tau:** Algol-type eclipsing binary 5.9–6.7 magnitude range; close binary system

12

Southern Skies

Many people who have enjoyed astronomy and the night sky for any period of time will tell you to be sure to see the Southern Hemisphere at least once in your lifetime. Going south of the equator opens up a whole new universe for you to observe. And many of the objects in the southern skies are simply spectacular through a pair of binoculars.

Binoculars make for the perfect southern skies observing tool. They are easy to pack, and with a smaller pair such as 7×50s, no mount is required. If you decide to take a larger pair of binoculars that needs some sort of mount, this is manageable even with today's strict luggage requirements. You will also find binoculars easy to use, especially when observing foreign skies, and they provide you with a nice view. As is always true, the darker the skies, the better; avoid city lights for the best views.

On my first trip south of the equator years ago, my first take—even after having spent many hours under the skies of a planetarium—was that I had landed on an alien planet! Constellations like Centaurus and the Southern Cross are easy to identify. The Milky Way takes on a whole new dimension, especially during our summer (the Southern Hemisphere's winter). You can spend several evenings exploring the beauty of the Large and Small Magellanic Clouds, companion galaxies to our own Milky Way that truly look like clouds. And then there are the not-so-easy-to-find southern constellations and the elusive South Celestial Pole; there is no bright star like Polaris to mark the spot.

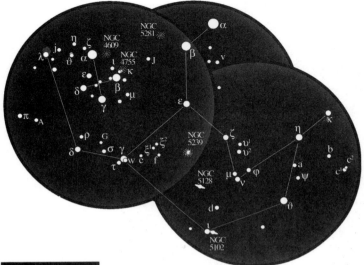

The beautiful Southern Celestial Sphere constellations of Crux, the Southern Cross, and Centaurus. Many evenings can be spent exploring this area of the sky with binoculars. Each circle is approximately 27 degrees in diameter. ILLUSTRATION BY DAVID FRANTZ

The southern skies take a bit of orientation for the first-time observer. I highly recommend purchasing a Southern Hemisphere star wheel or planisphere prior to your trip, even though I have found planispheres available in bookstores in some countries. You might also consider a star chart that will help you find more elusive objects as you star-hop around the southern skies.

Depending on how far south you travel, be prepared for strange sights like Orion—our most prominent Northern Hemisphere constellation—appearing to hang upside down in the sky. Even if you just cross the equator, you could catch our summer constellations such as Scorpius and Sagittarius overhead, instead of low on the southern horizon as we see them in the United States and other Northern Hemisphere locations.

The Astronomical League's Southern Skies Binocular List makes a great starting point as a reference for objects you can

Crux, the Southern Cross PHOTO BY VIC WINTER

observe. The list includes seventy-three objects readily observable with 7×50 binoculars. And there are regular star parties in the Southern Hemisphere. The most established is the Southern Skies Star Party on Lake Titicaca in Bolivia, at an altitude of 12,500 feet. On an annual basis, some forty astronomers of all types, from novices to seasoned observers, make the trek. The altitude takes a little getting used to, but the skies are simply a delight, by far the best I have ever observed. My friends David Levy, Vic Winter, and I marveled at the fact that the Milky Way was so bright that it cast a shadow. As we stood at a still Lake Titicaca, we could pick out the reflections of Messier Objects in the lake. If you are a novice, there are plenty of people who are more than willing to show you many of the Southern Hemisphere wonders through their binoculars and telescopes. It is an experience you will never forget.

SOUTHERN HEMISPHERE SELECT OBJECTS LIST

Each of the seasonal chapters has listed "low in the south" constellations and some of the terrific binocular objects. But if your time is limited, you should concentrate on a short list of objects that you most want to observe. There are a number of "don't miss" Southern Hemisphere deep-sky objects; these are all nice binocular objects and many of my personal favorites. They are listed below by time of year, starting in January at around 9:00 P.M. local time.

As with the Northern Hemisphere seasonal object listings, the objects outlined here are a good start. Spending time exploring the Milky Way—high in the sky during the Southern Hemisphere's winter months—is a treat in itself. There are numerous binaries, variables, and other objects to observe in the Southern Hemisphere skies. Take full advantage of these wonders, and you will be tempted to plan a repeat trip south of the equator.

47 Tucanae, NGC 104/C106. Better known simply as "47 Tuc," this globular cluster is a rival to Omega Centauri at magnitude 4. It is also visible to the naked eye and is similar in size to Omega Centauri, about the same apparent size as the moon. Some observers believe it is more spectacular in appearance than Omega Centauri.

Small Magellanic Cloud, NGC 292. Upon your first naked-eye glance, you will probably think clouds are moving in to spoil your viewing session. But the Small Magellanic Cloud (SMC) is an irreg-

47 Tucanae, NGC 104, above the Small Magellanic Cloud

PHOTO BY VIC WINTER

ular dwarf galaxy that is associated with our own Milky Way Galaxy. The SMC is a superb appetizer for the main treat: the Large Magellanic Cloud. The farther south you travel, the better view you will experience of both. The Small and Large Magellanic Clouds are satellite galaxies of the Milky Way.

Within the SMC are numerous globular clusters, open clusters, and open clusters associated with diffuse nebulae. Many of these objects are relatively faint, between tenth and twelfth magnitude. The brighter ones might just be visible in a pair of 50-millimeter binoculars, whereas a pair of 100-millimeter binoculars will show the fainter objects as well.

Large Magellanic Cloud. One word to describe the Large Magellanic Cloud (LMC): Wow! At a visual magnitude of 0.1, the LMC is a delight. The first preserved record was by the Persian astronomer Al Sufi, who, in his 964 A.D. *Book of Fixed Stars,* called it Al Bakr, or the White Ox. Others, including Amerigo Vespucci, noted the LMC, but it was Magellan and his expedition in 1519 that made it and the SMC well known. Like the SMC, the LMC is an irregular dwarf galaxy that is orbiting the Milky Way Galaxy.

Once you begin observing the LMC with a pair of binoculars, you won't want the sun to rise. The LMC contains numerous interesting objects, including diffuse nebulae, especially the Tarantula Nebula; globular and open clusters; planetary nebulae; and much

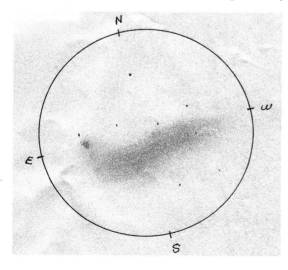

The Large Magellanic Cloud through 7×50s ILLUSTRATION BY GONZALO VARGAS

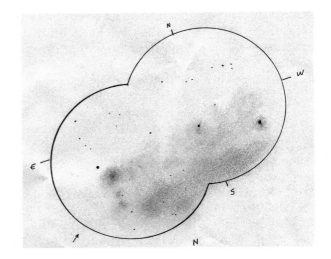

*The Tarantula
Nebula in the
Large Magel-
lanic Cloud*

ILLUSTRATION BY

GONZALO VARGAS

more. A few objects will be visible with 50-millimeter binoculars, but 100-millimeter binoculars will provide you with a spectacular view.

Tarantula Nebula, NGC 2070/C103, 30 Doradus. A diffuse nebula within the LMC, around magnitude 3 and fairly large, about 40 by 25 arc minutes. The Tarantula Nebula is also referred to as the True Lovers' Knot or Great Looped Nebula. The famous astronomer Sir John Herschel first called it the Looped Nebula. Supernova 1987A occurred on February 24, 1987, within the Tarantula Nebula. With a maximum brightness of around magnitude 2.9, it was the nearest observed supernova since Kepler's, which occurred before the telescope's invention.

NCG 2516/C96. A nice, large open cluster of blue-white stars at magnitude 3.8 and a diameter of 30 arc minutes, NGC 2516 was discovered by Abbe Lacaille from South Africa in 1751–52.

NGC 2547. An open cluster of blue-white stars at magnitude 4.7 and a size of 20 arc minutes.

IC 2395. A small magnitude 4.7 open cluster with a size of about 8 arc minutes.

NGC 2808. A magnitude 6.3, 13.8 arc minute globular cluster that exhibits a bright center.

12 Muscae. Also listed as NGC 4833/C105, 12 Muscae was discovered by Lacaille in 1751–52. This magnitude 7.3 globular cluster has a dimension of about 14 arc minutes.

NGC 3114. A nice magnitude 4 open cluster with a diameter a little larger than the size of the moon (35 arc minutes), with blue-white stars all at about the same magnitude.

IC 2602/C102. Often referred to as the Southern Pleiades, IC 2602 is a nice open cluster around magnitude 1.9 and about 50 arc minutes across. A terrific binocular object.

Eta Carina. At about magnitude 3 and 120 arc minutes in size, this giant diffuse nebula, NGC 3372/C92, has been observed and the subject of research over many years. The Eta Carina nebula is incredible in nebulous detail, with easily observable regions of dark and bright interstellar matter. The nebula itself includes a peculiar star, Eta Carinae, one of the most massive and luminous Milky Way stars. Within Eta Carina, look for the Keyhole Nebula, formed by Eta Carina's brightest area and dark material, and the Homunkulus Nebula, found surrounding the star Eta Carinae. You will also find a number of open clusters, including Bochum 10 and 11, Collinder 228, and Trumpler 14, 15, and 16. You can spend hours studying the beautiful, complex, and awe-inspiring Eta Carina. If you have the opportunity to travel to the Southern Hemisphere, where Eta Carina is high in the sky, this will become one of your favorite objects to visit and revisit.

IC 2602/C102. The Theta Carinae Cluster is a magnitude 1.9 open cluster 50 arc minutes in size. It is one of the finest open clusters, made up of about sixty stars.

NGC 3372, the spectacular Eta Carina

PHOTO BY VIC WINTER

NGC 3532/C91. A magnitude 3.0 open cluster, NGC 3532 is a very large object about 55 arc minutes in size, with blue-white stars.

NGC 3766/C97. A relatively small magnitude 5.3, 12 arc minute open cluster made up of bluish stars of mostly the same brightness.

NGC 4609/C98. A very small magnitude 6.9, 5 arc minute open cluster that appears fainter than magnitude 6.9. The rest of the field is fairly void of stars.

C99, Coalsack. A very large, dark nebula in the constellation Crux. You will be surprised at how dark and void of stars the field appears, even through a pair of binoculars. It is also large, roughly more than ten times the moon's diameter. The Coalsack is one object I can simply stare at for some time; it amazes me how dark this region is compared with the surrounding area. Astronomers understand that regions like the Coalsack are dense areas where star formation is taking place.

Jewel Box. Located near the Coalsack in the constellation of Crux, the Jewel Box, NGC 4755/C94, also known as Kappa Crucis

NGC 5139, Omega Centauri

PHOTO BY VIC WINTER

Cluster, is a beautiful open cluster of around magnitude 4.2 at about 10 arc minutes. The stars that make up the Jewel Box are hot, young stars.

Omega Centauri. The brightest globular cluster visible in the sky, also known as NGC 5139/C80, at magnitude 3.6 with a diameter of around 36 arc minutes. This beautiful object is visible even to the naked eye in dark skies. It appears to be the largest globular cluster associated with the Milky Way Galaxy. Omega Centauri is a favorite of southern U.S. observers. When observed from the Southern Hemisphere, it is even more spectacular because it is much higher in the sky. In a pair of 7×50s, expect to see a brighter center with no resolvable stars. I am constantly surprised at the beauty of Omega Centauri through binoculars, with its oval shape and almost granular texture.

Centaurus A. A galaxy of magnitude 7 and an apparent size of 18 by 14 arc minutes, Centaurus A will not look like a galaxy through a pair of binoculars. Discovered by James Dunlop in 1826, Centaurus A, or NGC 5128/C77, is a peculiar galaxy. ("Peculiar" is an actual classification of galaxy, like spiral or elliptical.) It is interesting to astronomers because it is a strong source of radio radiation and the closest such galaxy.

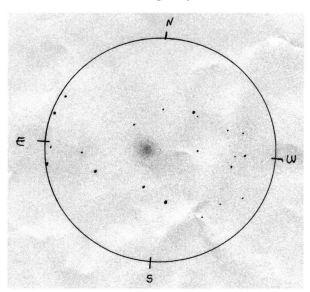

NGC 6397, a globular cluster in the constellation Ara

ILLUSTRATION BY
GONZALO VARGAS

Alpha Centauri. The brightest star in the constellation Centaurus, Alpha is a beautiful star in binoculars. It is in the closest star system to our solar system and sun. Alpha is a visual double star with a separation of 18.1 degrees.

NGC 6025/C95. This magnitude 5.1 open cluster with a size of 12 arc minutes is another object discovered by Abbe Lacaille. NGC 6025 includes around thirty stars of magnitude 7 or fainter.

NGC 6067. A magnitude 5.6, 12 arc minute open cluster with lots of stars, a few of them brighter. The stars appear bluish.

NGC 6087/C89. Another open cluster with bluish stars quite similar to NGC 6067, at magnitude 5.4 and also 12 arc minutes in size. NGC 6087 appears to have fewer stars than NGC 6067 and is Y-shaped or triangular.

NGC 6397/C86. A globular cluster that is magnitude 5.6 and about 26 arc minutes in size. NGC 6397 looks almost like an open cluster.

NGC 6752/C93. A big globular cluster with a bluish center, magnitude 5.4, and 20 arc minutes across.

Appendix A

References

A number of excellent references as well as observing guides and materials are available, from the novice to advanced level. Some of these are out of print, so you'll need to check with your local used bookstore or online for sources. A few classic books are included.

See appendix I for Planisphere sources as well as directions for use.

BINOCULARS AND EQUIPMENT

Hale, Alan R. *How To Choose Binoculars*. Redondo Beach, California: C & A Publishing, 1991.

Harrington, Philip. *Star Ware*. New York, New York: John Wiley & Sons, Inc., 2002.

Paul, Henry. *Binoculars and All Purpose Telescopes*. New York, New York: Amphoto, 1980.

Seyfried, J. W. *Choosing, Using and Repairing Binoculars*. Ann Arbor, Michigan: University Optics, Inc., 1995.

Watson, Fred. *Binoculars, Opera Glasses and Field Glasses*. Buckinghamshire, Great Britain: Shire Publications Ltd, 1995.

OBSERVING BOOKS AND GUIDES

The American Association of Variable Star Observers (AAVSO). *AAVSO Manual for Visual Observing of Variable Stars*. Cambridge, Massachusetts: The American Association of Variable Star Observers, 2001.

Ancient City Astronomy Club. *Observe The Herschel Objects*. Washington, D. C.: The Astronomical League, 1980.

Bakich, Michael E. *The Cambridge Encyclopedia of Amateur Astronomy.* New York, New York: Cambridge University Press, 2003.

Beckman, K. C. *Nova Hunter's Handbook.* Cambridge, Massachusetts: American Association of Variable Star Observers, 2000.

Bishop, Roy L., ed. *Observer's Handbook.* Toronto, Ontario, Canada: Royal Astronomical Society of Canada. Published annually.

Burnam, Robert et al. *Advanced Skywatching.* New York, New York: Time-Life Books, 1997.

Burnham, Robert Jr. *Burnham's Celestial Handbook.* 3 vols. New York, New York: Dover Publications, 1978.

Chandler, David. *Exploring the Night Sky with Binoculars.* La Verne, California: David Chandler, 1994.

Cherrington, Ernest. *Exploring the Moon Through Binoculars and Small Telescopes.* New York, New York: Dover Publications, 1984.

Crossen, Craig and Gerald Rhemann. *Sky Vistas: Astronomy for Binoculars and Richest-Field Telescopes.* New York, New York: Springer Wien, 2004.

Crossen, Craig and Wil Tirion. *Binocular Astronomy.* Richmond, Virginia: Willmann-Bell, Inc., 1992.

Dickinson, Terence. *The Backyard Astronomer's Guide.* Camden East, Ontario, Canada: Camden House, 1991.

———. *Nightwatch: An Equinox Guide.* Camden East, Ontario, Canada: Camden House, 1983.

———. *Nightwatch: A Practical Guide to Viewing the Universe.* Willowdale, Ontario, Canada: Firefly Books, 2001.

———. *Summer Stargazing.* Camden East, Ontario, Canada: Camden House, 1996.

Discovery Channel. *Night Sky.* New York, New York: Discovery Books, 1999.

Evans, Robert O. *AAVSO Supernova Search Manual.* Cambridge, Massachusetts: American Association of Variable Star Observers, 1993.

Goldberg, Amelia. *Universe Sampler: A Journey Through the Universe For the Beginner.* Washington, D. C.: The Astronomical League, 1998.

Harrington, Philip. *Touring the Universe through Binoculars.* New York, New York: John Wiley & Sons, Inc., 1990.

Heifetz, Milton and Wil Tirion. *A Walk through the Heavens: A Guide to Stars and Constellations and Their Legends.* Cambridge, United Kingdom: Cambridge University Press, 2002.

Hill, Richard E., ed. *Observe and Understand the Sun.* Washington, D. C.: The Astronomical League, 2000.

Hoffmeister, Cuno, G. Richter, and W. Wenzel. *Variable Stars.* New York, New York: Springer-Verlag, 1985.

Kolman, Roger S. *Observe and Understand Variable Stars.* Washington, D. C.: The Astronomical League, 1999.

Levitt, I. M. and Roy K. Marshall. *Star Maps for Beginners.* New York, New York: Simon and Schuster, 1987.

Levy, David H. *Observing Variable Stars: A Guide for the Beginner.* New York, New York: Cambridge University Press, 1989.

———. *The Sky, A User's Guide.* New York, New York: Cambridge University Press, 1993.

Lunsford, Robert D. *The A.L.P.O. Guide to Watching Meteors.* Washington, D. C.: The Astronomical League, 1995.

Machholz, Don. *Messier Marathon Observer's Guide.* Colfax, California: Make Wood Products, 1994.

———. *The Observing Guide to the Messier Marathon: A Handbook and Atlas.* New York, New York: Cambridge University Press, 2002.

Machin, Kathy and Sue Wheatley. *Messier Objects: A Beginner's Guide.* Washington, D. C.: The Astronomical League, 1997.

MacRobert, Alan. *Star Hopping for Backyard Astronomers.* Belmont, Massachusetts: Sky Publishing Corporation, 1994.

Mayall, Newton, Margaret Mayall, and Jerome Wyckoff. *The Sky Observer's Guide.* New York, New York: Golden Press, 1959.

Mechler, Gary. *First Field Guide Night Sky.* New York: Scholastic, Inc., 1999.

Mechler, Gary and Mark Chartrand. *National Audubon Society Pocket Guide to Constellations of the Northern Sky.* New York, New York: Albert Knopf, 1995.

Mensing, Stephen. *Star Gazing Through Binoculars.* Blue Ridge Summit, Pennsylvania: Tab Books, Inc., 1986.

Moore, Patrick. *Exploring the Night Sky with Binoculars.* Cambridge, United Kingdom: Cambridge University Press, 2000.

———. *Guide to the Night Sky.* London, United Kingdom: George Philip Limited, 2001.

————. *Stargazing: Astronomy Without a Telescope.* Cambridge, United Kingdom: Cambridge University Press, 2001.

Mosley, John. *Stargazing: Exploring the Stars with Binoculars & Telescopes.* New York, New York: Barnes & Noble Books, 2000.

Ottewell, Guy. *Astronomical Calendar.* Greenville, South Carolina: Universal Workshop, Furman University. Published annually.

Pasachoff, Jay M. *Peterson Field Guide to the Stars and Planets.* Boston, Massachusetts: Houghton Mifflin, 2000.

Peltier, Leslie C. *The Binocular Stargazer.* Waukesha, Wisconsin: Kalmbach Publishing Company, 1995.

Pennington, Harvard. *The Year-Round Messier Marathon Field Guide: With Complete Maps, Charts and Tips to Guide You to Enjoying the Most Famous List of Deep-Sky Objects.* Richmond, Virginia: Willmann-Bell, 1997.

Percy, John R., ed. *The Study of Variable Stars Using Small Telescopes.* New York, New York: Cambridge University Press, 1986.

Rey, H. (updated by Jay M. Pasachoff). *The Stars: A New Way To See Them.* Boston, Massachusetts: Houghton Mifflin, 1989.

Reynolds, Mike and Richard Sweetsir. *Observe: Eclipses.* 2nd ed. Washington, D. C.: The Astronomical League, 1995.

Ridpath, Ian and Wil Tirion. *The Monthly Sky Guide.* 5th ed. New York, New York: Cambridge University Press, 1999.

Schaaf, Fred. *40 Nights to Knowing the Sky.* New York, New York: Henry Holt and Company, 1998.

Serviss, Garrett P. *Astronomy With An Opera-Glass.* New York, New York: D. Appleton and Company, 1912.

Tanguay, Ronald Charles. *The Double Star Observer's Handbook.* 2nd ed. Saugus, Massachusetts: Double Star Observer, 2003.

Van Holt, Tom. *Stargazing.* Mechanicsburg, Pennsylvania: Stackpole Books, 1999.

Whitney, Charles A. *Whitney's Star Finder.* New York, New York: Random House, 1990.

Wood, Charles A. *The Modern Moon: A Personal View.* Cambridge, Massachusetts: Sky Publishing Corporation, 2003.

Zim, Herbert and Robert Baker. *Stars: A Guide to the Constellations, Sun, Moon, Planets, and Other Features of the Heavens.* New York, New York: St. Martin's Press, 1985.

GENERAL ASTRONOMY

Allen, Richard Hinckley. *Star Names: Their Lore and Meaning.* New York, New York: Dover, 1963.

Chaisson, Eric and Steve McMillan. *Astronomy: A Beginner's Guide to the Universe.* Englewood Cliffs, New Jersey: Prentice Hall, 1995.

———. *Astronomy Today.* Upper Saddle River, New Jersey: Prentice Hall, 1997.

Dickinson, Terence. *The Universe and Beyond.* Camden East, Ontario, Canada: Camden House, 1992.

Dickinson, Terence and Jack Newton. *Splendors of the Universe.* Camden East, Ontario, Canada: Camden House, 1997.

Ferris, Timothy. *Coming of Age in the Milky Way.* New York, New York: Morrow, 1988.

Forey, Pamela and Cecilia Fitzsimons. *An Instant Guide to Stars & Planets.* New York, New York: Gramercy Books, 1999.

Gribbin, John and Simon Goodwin. *Origins: Our Place in Hubble's Universe.* Woodstock, New York: Overlook Press, 1998.

Jones, Brian. *The Practical Astronomer.* New York, New York: Simon & Schuster, 1990.

Kaler, James B. *The Ever-Changing Sky: A Guide to the Celestial Sphere.* New York, New York: Cambridge University Press, 1996.

Malin, David. *A View of the Universe.* New York, New York: Cambridge University Press, 1993.

Moche, Dinah L. *Astronomy: A Self-Teaching Guide.* New York, New York: Wiley, 1993.

Moore, Patrick and Wil Tirion. *Cambridge Guide to Stars and Planets.* Cambridge, United Kingdom: Cambridge University Press, 2000.

Pasachoff, Jay M. *Astronomy from the Earth to the Universe.* 5th ed. Philadelphia, Pennsylvania: Saunders, 1997.

Percy, John R., Janet Akyüz Mattei, and Christiaan Sterken, eds. *Variable Star Research: An International Perspective.* New York, New York: Cambridge University Press, 1992.

Reynolds, Mike. *Falling Stars: A Guide to Meteors and Meteorites.* Mechanicsburg, Pennsylvania: Stackpole Books, 2001.

STAR ATLASES

Arnold, HJP, Paul Doherty, and Patrick Moore. *The Photographic Atlas of the Stars*. Philadelphia, Pennsylvania: Institute of Physics Publishing, 1999. [Goes down to approximately magnitude 8]

Ridpath, Ian, ed. *Norton's Star Atlas and Reference Handbook*. New York: Pearson, 2004. [Goes down to magnitude 6]

Scovil, Charles, ed. *AAVSO Variable Star Atlas*. Cambridge, Massachusetts: American Association of Variable Star Observers, 1990. [Goes down to magnitude 9.5]

Sinnott, Roger W. and Michael A. C. Perryman. *Millennium Star Atlas*. Cambridge, Massachusetts: Sky Publishing Corporation, 1997. [Goes down to magnitude 11]

Tirion, Wil. *Cambridge Star Atlas*. 2nd ed. New York, New York: Cambridge University Press, 1996. [Goes down to magnitude 6]

Tirion, Wil and Roger W. Sinnott. *Sky Atlas 2000.0* 2nd ed. Cambridge, Massachusetts: Sky Publishing Corporation, 1998. [Goes down to magnitude 8.5]

Tirion, Wil, Barry Rappaport, and George Lovi. *Uranometria 2000.0*. Vol. 1, Northern Hemisphere; Vol. 2, Southern Hemisphere. Richmond, Virginia: Willmann-Bell, 1993.[Goes down to magnitude 9+]

SOFTWARE

MegaStar.
  Willmann-Bell, Richmond, Virginia (www.willbell.com)
Red Shift.
  Maris Multimedia, Ltd., Kingston, United Kingdom (www.maris.com)
Starry Night Backyard and Starry Night Pro.
  Sienna Software, Toronto, Ontario, Canada (www.siennasoft.com)
TheSky and *RealSky.*
  Software Bisque, Golden, Colorado (www.bisque.com)
Touring the Universe Through Binoculars Star Atlas.
  http://philharrington.net

PERIODICALS
The following periodicals overview events in the night sky, as well as provide observing hints, activities, and report observations:

Journal of the American Association of Variable Star Observers
AAVSO
25 Birch Street
Cambridge, MA 02138
e-mail: aavso@aavso.org
www.aavso.org

Astronomical Calendar
(Annual publication by Guy Ottewell)
Astronomical Workshop
Furman University
Greenville, SC 29613
Order the *Astronomical Calendar* at
http://SkyandTelescope.com/astronomicalcalendar
or through your local astronomy club

The Reflector
The Astronomical League
11305 King Street
Overland Park, KS 66210-3421
www.astroleague.org

Astronomy
21027 Crossroads Circle
Waukesha, WI 53187
(800) 446-5489
e-mail: customerservice@kalmbach.com
www.kalmback.com/astro/astronomy.html
Features a nice monthly binocular astronomy column entitled "Phil Harrington's Binocular Universe" that is well-suited for beginners.

Mercury
Astronomical Society of the Pacific
390 Ashton Avenue
San Francisco, CA 94112

e-mail: newsletter@astrosociety.org
www.astrosociety.org

Journal of the British Astronomical Association
The British Astronomical Association
Burlington House
Piccadilly
London W1V 9AG, England
e-mail: office@britastro.com
www.britastro.org

Night Sky
49 Bay State Road
Cambridge, MA 02138-9922
www.nightskymag.com

The Planetary Report
The Planetary Society
65 North Catalina Avenue
Pasadena, CA 91106
www.planetary.org

Observer's Handbook
(Annual publication)
The Royal Astronomical Society of Canada
136 Dupont Street
Toronto, Ontario M5R 1V2, Canada
e-mail: nationaloffice@rasc.ca
mempub@rasc.ca
www.rasc.ca

Sky & Telescope
Post Office Box 9111
Belmont, MA 02178-9111
(800) 253-0245
e-mail: orders@skypub.com
www.skypub.com
Has a pair of helpful binocular astronomy columns that are good for beginners: "Northern and Southern Binocular Highlights."

The Journal of the Association of Lunar and Planetary Observers
The Strolling Astronomer
  The Association of Lunar and Planetary Astronomers
  Post Office Box 13456
  Springfield, IL 62791-3456
  www.lpl.arizona.edu/alpo/

WEBSITES

Peter Abrahams's site includes extensive research about the history of telescope and binoculars. Abrahams, a former President of the Antique Telescope Society, is an avid historian.
  http://home.europa.com/~telscope/binotele.htm

www.astronomy.net: A website with a very wide variety of astronomy resources, including manufacturers and distributors of astronomical equipment, local astronomy clubs, events, observing, and other threads that one might find interesting or of use.

Astronomy Picture of the Day: Perhaps one of the most-hit sites on the web today, Astronomy Picture of the Day, often referred to as APOD, is originated, written, coordinated, and edited by Robert Nemiroff and Jerry Bonnell since 1995. The APOD archive contains the largest collection of annotated astronomical images on the internet.
  http://antwrp.gsfc.nasa.gov/apod/astropix.html

David Paul Green site gives readers an opportunity to download shareware programs and documents including The Ultimate Messier Object List, Overlooked Object Log, and The Caldwell Observer's Log.
  www.davidpaulgreen.com

Appendix B

Manufacturers and Retailers

The following list of dealers and sources is not meant to be exhaustive, nor is it meant to provide recommendations. It is simply a list with which you can begin doing research on binoculars, accessories, and telescope binocular viewers. A number of distributors for manufacturers like Celestron and Meade are also noted.

BINOCULAR MANUFACTURERS AND DEALERS
Adorama
One of the names familiar to those buying cameras and camera accessories mail order, Adorama also carries a number of pairs of binoculars, including Bushnell, Canon, Celestron, Fujinon, Nikon, and Zeiss.

42 West 18th Street
New York, NY 10011
Orders: (800) 223-2500
Information: (212) 741-0052
e-mail: info@adorama.com
www.adorama.com

Anacortes Telescope and Wild Bird
A major source of astronomical equipment, Anacortes carries numerous brands and models, including Burgess, Canon, Fujinon, Leica, Minox, Miyauchi, Nikon, Pentax, Takahashi, and Zeiss. They also carry various binocular mount systems, including Universal Astronomics UniMounts. Additionally, you will find the Celestron

and Tele Vue telescope binocular viewers and the JMI Reverse
Binoculars.

9973 Padilla Heights Road
Anacortes, WA 98221
Orders: (800) 850-2001
Information: (360) 588-9000
Fax: (360) 588-9100
e-mail: info@buytelescopes.com
www. buytelescopes.com

Apogee, Inc.

A distributor of several telescope brands and models, Apogee also
sells their own brand of binoculars as well as Galileo. A number of
options are available, including the Astro-Vue series with built-in,
flip in-flip out dark sky filters and the right-angle model RA-88-SA.

Post Office Box 136
Union, Il 60180
Orders: (877) 923-1602
Information: (815) 568-2880
Fax: (815) 568-1998
e-mail: apogeeinc@sbcglobal.net
www.apogeeinc.com

AstroMart

A website for buying and selling items, AstroMart allows registered
participants the opportunity to connect to buy (or sell) an item. The
seller and buyer negotiate any and all final details, including form
of payment. Usually good deals can be found, but as with all sites,
always use common sense.

www.astromart.com

Astronomics

An established major distributor of telescopes and accessories,
Astronomics also carries over fifteen brands of binoculars, including Bushnell, Canon, Fujinon, Meade, Takahashi, and Zeiss.

680 SW 24th Avenue
Norman, OK 73069
Orders: (800) 422-7876
Information: (405) 364-0858

Fax: (405) 447-3337
e-mail: questions@astronomics.com
www.astronomics.com

BigBinoculars.com

A mail-order Internet site, BigBinoculars.com features Fujinon, JMI Reverse Binoculars, Miyauchi, and Oberwerk, as well as some less-known brands like the Italian-made Astromeccanica refracting telescope binocular and Kronos. They also carry a number of binocular mounts and mounting systems, including Blaho and Universal Astronomics.

Orders: (866) 244-2460
Information: (937) 372-5409
Fax: (937) 376-7903
www.bigbinoculars.com

Burgess Optical Company

A mail-order and Internet site, Burgess Optical offers its own line of binoculars, telescopes, and binocular mounts. Other accessories, like telescope binocular viewers and eyepieces, are also available. Binocular models include six models, from 8×42s and 10×42s (roof prisms) to giants in 70, 80, and 100 millimeter diameters. Burgess binoculars are also carried by other dealers.

7756 Oak Ridge Highway
Knoxville, TN 37931
(865) 769-8777
e-mail: burgoptics@aol.com
www.burgessoptical.com

Camera Concepts and Telescope Solutions

In addition to two stores, Camera Concepts offers a large selection of items through their website. Included are several binocular brands like Celestron, Meade, Minolta, Nikon, and Oberwerk, Universal Astronomics mounts, and Celestron, Denkmeier, and Tele Vue telescope binocular viewers.

32 East Main Street
Patchogue, NY 11772
e-mail: camcon@msn.com
www.camercaconcepts.com

Coronado

Quickly becoming the source for solar astronomy, Coronado has developed a number of products for white light and hydrogen-alpha (H-a) observing. These products include white light and hydrogen-alpha binoculars. A number of dealers also stock Coronado products.

1674 S. Research Loop
Suite 436
Tucson, AZ 85710
(866) SUNWATCH
(520) 740-1561
Fax: (520) 624-5083
e-mail: info@coronadofilters.com
www.coronadofilters.com

Efston Science

A Canadian company, they feature the Japanese Carton Adlerblick binocular series.

3350 Dufferin Street
Toronto, Ontario M6A 3A4
Canada
Orders: (888) 777-5255
Information: (416) 787-5140
Fax: (888) 681-1079
e-mail: info@e-sci.com
www.telescopes.ca

Focus Camera

Another of the names familiar to those buying cameras mail order, Focus Camera also stocks telescopes (Celestron and Meade) as well as binoculars. Brands include Fujinon, Leica, Nikon, Steiner, and Zeiss.

905 McDonald Avenue
Brooklyn, NY 11218
Orders: (800) 221-0828
Information and New York residents: (718) 437-8810
Fax: (877) 863-6288
www.focuscamera.com

Hands On Optics

A distributor of a variety of equipment. Hands On Optics carries Celestron and Meade binoculars. They also carry the Denkmeier telescope binocular viewers. You might also want to check their used or close outs; they usually have a good selection available, occasionally including binocular mounts.

Box 10025
Gaithersburg, MD 20898
Orders: (866) SCOPES1
Information: (301) 482-0000
Fax: (301) 482-2210
e-mail: astroguy@handsonoptics.com
www.handsonoptics.com

Jim's Mobile Inc.

A company that has always produced innovative products, JMI has recently introduced their own line of reverse reflector telescope binoculars in 6-inch and 10-inch primary mirrors. As noted, a few distributors also carry the JMI RB Line of reverse telescope binoculars.

810 Quail Street, Unit E
Lakewood, CO 80215
Orders: (800) 247-0304
Information: (303) 233-5353
Fax: (303) 233-5359
www.jimsmobile.com

Kahn Scope Centre

For our fellow astronomers to the north as well as mail order, Kahn Scope Centre offers a few binoculars from manufacturers like Celestron, National Geographic, and the Coronado solar white light binoculars. They also carry used equipment.

3243 Dufferin Street
Toronto, Ontario M6A 2T2
Canada
Orders: (800) 580-7160
Information: (416) 783-4140
Fax: (416) 783-7697
e-mail: info@khanscope.com
www.khanscope.com

Island Eyepiece and Telescope Ltd.

Located in British Columbia, Island Eyepiece and Telescope stocks Antares (Canadian-made), Celestron, Omcon (Japanese-made), and Orion binoculars, as well as tripod adapters. They also carry three different manufacturers of telescope binocular viewers: Celestron, Denkmeier, and Tele Vue.

Post Office Box 133
Mill Bay, British Columbia V0R 2P0
Canada
(250) 743-6633
e-mail: sales@islandeyepiece.com
www.islandeyepiece.com

Kendrick Astro Instruments

Product lines include Kendrick's own line of dew removers and dew caps (telescopes), as well as solar filters specifically for binoculars. They are also distributors for a number of brands of binoculars, the Burgess binocular mounts, and the Denkmeier telescope binocular viewers. A number of distributors also carry the Kendrick products.

2920 Dundas St. West
Toronto, Ontario M6P 1Y8
Canada
(800) 393-5456
(416) 762-7946
Fax: (416) 762-2765
e-mail: Kendrick@kendrick-ai.com
www.kendrick-ai.com

New Mexico Astronomical

Carries Celestron, Meade, and Parks binoculars.

834 Gabaldon Road
Belen, NM 87002
(505) 720-5666
(505) 864-2953
e-mail: nmastronomical@aol.com
www. nmastronomical.com

Oberwerk Corporation

Making a big splash in binoculars, both for low prices and quality optics, Oberwerk binoculars are no longer a best-kept secret. Oberwerk offers binoculars in everything from 8×42 roof prisms (mostly for birders, hunters, and sporting events) to their large 100 mm astronomical binoculars, and a number of accessories, like tripods and mirror mounts.

> 2440 Wildwood
> Xenia, OH 45385
> (866) OBERWERK
> Fax: (937) 376-7903
> e-mail: info@oberwerk.com
> www.oberwerk.com

Oceanside Photo and Telescope

A well-known and established Southern California storefront (in which one could spend days!) as well as an online store which offers a wide variety of astronomical telescopes and accessories including binoculars and accessories such as Apogee, Bushnell, Canon, Celestron, Coronado BinoMite, JMI Reverse Binoculars, Konica Minolta, Leica, Meade, Nikon, NovaVue, Orion, Pentax, and Vixon.

> 918 Mission Avenue
> Oceanside, CA 92054
> (800) 48-FOCUS
> Fax: (760) 722-8133
> e-mail: opt@optcorp.com
> www.optcorp.com

OpticsPlanet.com

An online store offering a large selection of binoculars as well as general information. Binoculars include brands like Bushnell, Canon, Meade, Nikon, Oberwerk, and Tasco.

> www.opticsplanet.com

Orion Telescopes and Binoculars

A major source of telescopes and binoculars, with their own Orion brand of binoculars, as well as Bausch & Lomb, Celestron, Fujinon,

Nikon, and Swift (select models). Somewhere around thirty different pairs of binoculars are offered, including Orion's own 80 mm binocular telescope. Also available are accessories including binocular mounts and tripods. Orion has two storefronts in Cupertino and Watsonville, California.

Order: (800) 447-1001
Information and advice: (800) 676-1343
www.telescope.com

Rivers Camera

In addition to two storefronts, Rivers Camera offers orders via mail. They carry binoculars by manufacturers like Celestron, Leica, Meade, Minolta, Nikon, and Zeiss. In addition, Rivers is a Tele Vue dealer.

(800) 245-7963
454 Central Avenue
Dover, NH 03820
(603) 742-4888
69 North Main Street
Rochester, NH 03867
(603) 332-5652
e-mail: rivers@ttlc.net
www.riverscamera.com

Scope City

With a number of California (five) and one Nevada storefronts, Scope City also offers orders via mail. With an extensive selection of binocular manufacturers—26 different manufacturers like Barska, Canon, Celestron, Fujinon, Meade, Nikon, Parks, and Zeiss—one can find a wide variety of models and prices. Scope City also carries Celestron telescope binocular viewers and the JMI Reverse Binoculars. The Scope City storefronts allow you to "take a look" and compare before you buy. Their website is also extensive.

(800) 235-3344 (National)
(800) 847-2673 (California)
e-mail: sales@scopecity.com
www.scopecity.com

Shultan Camera & Video

Offering a large range of cameras, telescope, and accessories, Shultan also offers a number of binocular models from manufacturers including Bushnell, Bausch & Lomb, Canon, Celestron, Fujinon, Meade, Nikon, Swarovski, Swift, and Zeiss. In addition, they also sell the Celestron telescope binocular viewers.

100 Fairway Drive
Vernon Hills, IL 60061
(800) 621-2248
Illinois: (847) 367-4600
Fax: (847) 367-6611
e-mail: info@shutan.com
www.shultan.com

T & T Binocular Mounts, L.L.C.

Offering a number of different models of mounts for binoculars, including wheelchairs, T & T even has a crutch tripod available.

18 Strong Street Extension
East Haven, CT 06512
(203) 469-2845 or (203) 272-1915
e-mail: telloyd@aol.com
www.benjaminsweb.com/TandT/

Universal Astronomics

Considered by many to be the leader in binocular mounts, Universal Astronomics offers a number of mounts, tripods, and options to securely mount your binoculars. They also carry binoculars—Fujinon, Miyauchi, Oberwerk, and Takahashi—as well as telescopes, mounts, and other accessories.

6 River Court
Webster, MA 01570
(509) 943-5105
e-mail: larry@universalastronomics.com
www.universalastronomics.com

Wolf Camera

Part of the Ritz Camera firm, Wolf Camera stores are often found in local malls. One specific Wolf Camera in Sarasota, Florida, has an astronomer on staff, as well as a large selection of binoculars including Celestron, Meade, Fujinon, Minolta, Nikon, Pentax, Swarovski, and Zeiss, all of which can be mail ordered.

2069 Siesta Drive
Sarasota, FL 34239
(941) 955-3537
Fax: (941) 366-8640
e-mail: cpisa@mindspring.com
www.ritzcamera.com

Woodland Hills Camera and Telescopes

A distributor for a number of binocular brands, as well as telescopes and accessories, Woodland Hills stocks Bushnell, Bausch & Lomb, Canon, Celestron, Leica, Minolta, Nikon, Pentax, Swarovski, and Zeiss binoculars.

5348 Topanga Canyon Boulevard
Woodland Hills, CA 91364
Orders: (888) 427-8766 (888-4astronomy)
Information: (818) 347-2270
Fax: (818) 992-4486
www.telescopes.net

ACCESSORIES

Blaho Company

Makers of binocular mounts, their website will overview their mounts and accessories, as well as direct you to the dealers who carry their products.

e-mail: blahocompany@enigma22.com
www.blaho.com

Denkmeier Optical Inc.

Manufacturers of telescope binocular viewers, Denkmeier's fine products include two lines with various accessories depending on the telescope on which the binoviewer will be used. A number of dealers also carry the Denkmeier binoviewers.

12636 Sunset Avenue J2
Ocean City, MD 21842
Orders: (866) 340-4578
Information: (410) 208-6014

Fax: (410) 208-6505
e-mail: deepskybinoviewer@deepskybinoviewer.com
www.denkmeier.com

Lumicon

Renowned for their filters, they also have their own telescope
binocular viewers (uses BK-7 glass in the Porro prisms); a number
of dealers stock Lumicon products.

750 Easy Street
Simi Valley, CA 93065
(805) 520-0047
Fax: (805) 520-3030
www.lumicon.com

Tele Vue Optics, Inc.

Known for their quality optics, Tele Vue also manufactures their
own telescope binocular viewer, Bino Vue. There are also a number
of accessories available. Worldwide you will find numerous Tele
Vue dealers.

32 Elkay Drive
Chester, NY 10918
(845) 469-4551
www.televue.com

Thousand Oaks Optical

A long-time solar filter producer with a wide variety of types and
sizes of solar filters, eclipse viewers, and nebular filters.

Box 4813
Thousand Oaks, CA 91359
Orders: (800) 996-9111
Information: (805) 491-3642
Fax: (805) 491-2393
www.thousandoaksoptical.com

Trico Machine Products

Manufacturers of the Sky Window, a binocular mount that allows
the observer to look down into a mirror. Trico also has a variety of
accessories available, as well as a couple of Sky Window–binocular
combinations.

5081 Corbin Drive
Bedford Heights, OH 44021
(216) 662-4194
Fax: (216) 662-7513
e-mail: info@tricomachine.com
www.tricomachine.com/skywindow/

University Optics, Inc.
A company that has been around for a number of years, they offer their own telescope binocular viewers that feature a 45-degree angle of viewing.

Post Office Box 1205
Ann Arbor, MI 48106
(734) 665-3575
(800) 521-2828
Fax: (734) 665-1815
e-mail: uoptics@aol.com
www.universityoptics.com

Appendix C

Observing Programs

There are a number of excellent organizations that promote observing of various types. Below is an overview of several U.S.-based (or, in some cases, simply English-language) organizations, including a binocular astronomy listserv. This alphabetically ordered list is not meant to be exclusive but to serve as a starting point for the new binocular astronomy observer.

THE AMERICAN ASSOCIATION
OF VARIABLE STAR OBSERVERS

The American Association of Variable Star Observers (AAVSO) was founded in 1911 at Harvard College Observatory to coordinate variable star observations made primarily by amateur astronomers. AAVSO membership is open to observers interested in variable stars who want to contribute research data. Approximately 650 observers from about forty countries participate each year in AAVSO observing programs, contributing around 300,000 observations each year. Observers are of all ages and levels of experience. Their equipment varies from the unaided eye to binoculars to exotic telescopes and CCD cameras. New observations are added to the data files for each star and the corresponding computer-generated light curves are then updated.

AAVSO members and observers meet each spring and fall to share observations and exchange ideas. The AAVSO also has a number of publications that contain valuable observing information. Observations of most variable stars types are coordinated and published by the AAVSO.

Although the procedures followed by AAVSO observers in the observing programs are standardized, each observer chooses his or her own areas of interest. Observers may choose one or several observing programs in which to participate. An observer may make one observation or thousands in a year, and may concentrate on one type of variable, many types of variables, the sun, and/or novae and supernovae.

Visual observations of variable stars comprise the largest number of observations: in the AAVSO visual observing program, one will find approximately five thousand objects. Most of these stars undergo visual changes greater than one magnitude. The types of variable stars include pulsating, eruptive or cataclysmic, nebular, irregular, and suspected variable stars. Contributions to the visual observing program are also made from other non-visual techniques. CCD or photoelectric photometry instruments can be used to record data, which complements visual observations.

There are a number of other AAVSO programs, including eclipsing binary stars, RR Lyrae stars, sunspot observations, indirect detection of solar flares, visual Nova Search, and a visual Supernova Search. The AAVSO website is outstanding and gives an excellent overview of its programs, available data and materials, and so on.

To contact the AAVSO:
The American Association of Variable Star Observers
25 Birch Street
Cambridge, MA 02138
(617) 354-0484
Fax: (617) 354-0665
e-mail: aavso@aavso.org
www.aavso.org

THE ASSOCIATION OF LUNAR AND PLANETARY OBSERVERS

The Association of Lunar and Planetary Observers (ALPO) was founded by Walter H. Haas in 1947 as a conduit for advancing and conducting astronomical work by both professional and amateur astronomers who share an interest in solar system observations. For the novice observer, the ALPO is a place to learn and to enhance observational techniques. For the advanced amateur astronomer, it is an opportunity to contribute data. For the professional astronomer, it is a resource where group studies or systematic observing patrols add to the advancement of astronomy.

Most solar system observing requires a telescope. However, there are some specific observations binocular users can make and contribute—for example, telescopic/binocular meteor observing, some lunar, comet observations, and eclipses

As the organization's name implies, members concentrates their observations on the sun, moon, planets, asteroids, meteors, and comets. The goals are to stimulate, coordinate, and generally promote the study of these solar system bodies using methods and instruments that are available within the communities of both amateur and professional astronomers. ALPO holds a conference each summer, usually in conjunction with other astronomical groups like the Astronomical League.

ALPO has "sections" for the observation of all the types of bodies found in our solar system. Section coordinators collect and study submitted observations, correspond with observers, encourage beginners, and contribute reports to our Journal at appropriate intervals. Each coordinator can supply observing forms and other instructional material to assist in your observations. You are encouraged to correspond with the coordinators in whose projects you are interested. Coordinators can be contacted through the ALPO website via e-mail or at their postal mail addresses listed in back of the ALPO Journal. ALPO member activities are on a volunteer basis, and each member can do as much or as little as he or she wishes.

To contact the ALPO:
  The Association of Lunar and Planetary Observers
  Post Office Box 13456
  Springfield, IL 62791-3456
  www.lpl.arizona.edu/alpo

THE ASTRONOMICAL LEAGUE'S OBSERVING PROGRAM

The Astronomical League is composed of over two hundred local amateur astronomy clubs and societies from all across the United

States. These organizations, along with Members-at-Large, Patrons, and Supporting members, form one of the largest amateur astronomical organizations in the world.

The major benefit of belonging to this organization is receiving the quarterly newsletter, *The Reflector,* which keeps you in touch with amateur astronomy activities all over the country. The chance to meet the people you read about there occurs during the League's annual National Convention, or at one of the ten regional conventions that the Astronomical League sponsors. The easiest way to become part of the League is to join one of the League's member societies or clubs. A benefit of membership in a local club or society is membership in the Astronomical League and part of your society dues goes to pay for your *Reflector* subscription. If it is not convenient for you to join one of these societies, you might consider becoming a Member-at-Large.

The Astronomical League has established a number of Observing Clubs, three of which are overviewed below. These Observing Clubs offer an organized system for recording your observations of objects. As you meet certain observing qualifications, you are rewarded with a certificate and pin for your efforts. To qualify for any of the League's observing awards listed below, you will need to be a member of the Astronomical League through either an affiliated society or as a Member-at-Large, and observe and record the number of objects required for each award.

Binocular Messier Club

The Binocular Messier Club is for beginning observers as well as experienced amateur astronomers. Beginning observers will find that it takes only a pair of binoculars, no matter what size, cost, or condition, to do enjoyable astronomy. On the other hand, many experienced amateurs, even though they may already have the League's telescopic Messier and Herschel certificates, will find that they enjoy the new perspective binocular observing provides.

To qualify for the League's Binocular Messier Certificate, you need to be a member of the League and observe 50 or more Messier objects using only binoculars. Any 50 of the 110 recognized Messier objects may be observed. Any pair of binoculars may be used, but those with objectives between 20 mm and 80 mm in diameter are recommended. To record your observations, you may use the log sheets found in the back of the Astronomical League's manual *Observe: A Guide to the Messier Objects*, as found in *Binocular Stargazing*, or any similar log sheet. The minimum required information includes the name of the object; date and time of the observation; an estimate of the seeing and transparency; the size and power of the binoculars used; and perhaps, a brief comment on what you saw.

Deep Sky Binocular Club

The Deep Sky Binocular Club is a list of sixty selected non-Messier objects and picks up where the Binocular Messier Club leaves off. The purpose of the Deep Sky Binocular Club is not to put your observing skills to the test by including the toughest objects observable with binoculars, but to allow you to observe and enjoy sixty of the most beautiful objects in the heavens; objects other than those discovered by Charles Messier.

Just because the Deep Sky Club comes after the Binocular Messier Club, doesn't mean you have to do your Messier observations before your deep sky observations. However, the Astronomical League recommends that you get your Binocular Messier Certificate first, before the Deep Sky Certificate, since Messier cataloged most of the easy objects. Even though the sixty objects in the Deep Sky Club are the best objects for small binoculars, it doesn't mean that they are all easy. For some of the objects, you will have to go to a good dark sky site, on a clear night with good seeing, and then observe those objects at the meridian for best results. All of the objects listed in the Deep Sky Club were observed with 7×50 Orion Explorer binoculars. For northern observers, no object on the list is below minus thirty-five degrees declination, which is the declination of the most southerly Messier object, M7.

To qualify for the Astronomical League's Deep Sky Binocular Certificate you need to be a member of the League and observe the sixty selected objects using binoculars. Any pair of binoculars may be used, but those with objectives between 50mm and 80mm in diameter are recommended. To record your observations, again you may use log sheets as noted above. There are also several rules to follow to obtain an award; please consult the Astronomical League's website to see updated rules.

Southern Skies Binocular Club

The third binocular observing program of the Astronomical League is its Southern Skies Binocular Club. The Binocular Messier Club in tandem with the Deep Sky Binocular Club presents a complete list of binocular objects that can be reasonably observed in the Northern Hemisphere. Likewise, the Southern Skies Binocular Club presents a complete list of binocular objects that can be reasonably observed from the Southern Hemisphere. Together, this trilogy of binocular clubs gives one an all-sky survey of deep sky objects that are within range of small aperture binoculars.

Over two hundred Southern Hemisphere deep sky objects were surveyed to come up with this list of seventy-three objects readily observable with 7×50 binoculars. All objects on this list are below minus forty degrees declination to qualify as a Southern Hemisphere object. To qualify for the Astronomical League's Southern Skies Binocular Club Certificate, you need observe fifty of the seventy-three objects on the list. You may choose which fifty objects you want to observe and submit for your certificate. The reason for this is that many of us will only have limited access to the Southern Hemisphere, maybe only one or two visits in a lifetime, while, at the same time, one may not have control over the best time of the year to observe in the Southern Hemisphere. An experienced binocular observer with two good nights of observing any time during the year should be able to acquire this certificate.

To qualify for the League's Southern Skies Binocular Certificate, you need to be a member of the League and observe fifty objects on the list using binoculars. Any pair of binoculars may be used, but those with objectives between 50 mm and 80 mm in diameter are recommended. To record your observations, again use a log as outlined.

To receive your observing certificate(s) for these Observing Clubs, send your observations along with your name, address, phone number, and club/society affiliation to:

Mike Benson
2308 Dundee Lane
Nashville, TN 37214-1520
e-mail: ocentaurus@aol.com
www.astroleague.org/observing.html

Lunar Club

Even though the League's Lunar Club includes visual and telescopic features on the moon, in addition to binocular features, it is included here as an excellent reference to preparing a list of lunar objects visible through binoculars. The Lunar Club list includes 100 features: 18 naked eye, 46 binocular, and 36 telescope. These objects have all been observed with 7×35 binoculars and a 60 mm refracting telescope.

This is an excellent project if you live in a city where light pollution is an issue or for schools and students. And since the moon interferes with observing deep sky objects, this is another opportunity to observe and learn the sky—specifically the moon—while awaiting the dark skies to observe deep-sky objects.

To receive your Lunar Club Certificate, send your observations along with your name, address, phone number, and club/society affiliation to:

Steve A. Nathan
AL Lunar Club Coordinator
45 Brewster Road
West Springfield, MA 01089
e-mail: snathan@k12.oit.umass.edu
www.astroleague.org/al/obsclubs/lunar/lunar1.html

For information on the AL:

The Astronomical League
9201 Ward Parkway, Suite #100
Kansas City, MO 64114
(816) 333-7759 (816 Deep Sky)
e-mail: aloffice@earthlink.net
www.astroleague.org

BINOCULAR ASTRONOMY YAHOO! GROUP

Among its many special interest groups, Yahoo! supports a Binocular Astronomy Group. This excellent listserv promotes threads of everything from equipment use and reviews, useful books and publications especially for the binocular astronomy enthusiast, to challenging binocular objects. The number of daily messages is not overwhelming and will provide the beginner with a wide range of advice.

Even though the Yahoo! Binocular Astronomy Group is not an official organization like others listed in this appendix, it provides an excellent service and useful resource to all interested in binocular astronomy. Group members are eager to provide sound advice to each other as well as promote the wonders of binocular astronomy.

For information:
 http://groups.yahoo.com/group/binocularastronomy/

DOUBLE STARS

Few organizations concentrate solely on observing double stars. A group based at the University of South Alabama recently organized and began publishing *The Journal of Double Star Observations*. This international amateur/professional journal is published as an e-zine: offered only in electronic format. Therefore, it is free of charge; one just needs to download each issue. For a number of years prior to the Journal, Ron Tanguay provided a subscription-based publication, titled *Double Star Observer*, which is no longer available.

The Journal's April 2005 inaugural issue notes that it has a technical slant, but the Journal also has the goal of publishing "fun" articles that showcase the pursuit of double stars and its excitement. Future articles will include "my favorite double," tips for observing, and profiles of other double star observers.

Most double star observations are made with high quality, high resolution telescopes with diameters of three inches or larger. Higher magnifications are often used. However, there are numerous double stars which make for excellent binocular objects. If you decide to graduate from binoculars to a telescope and enjoy observing double stars, you should contact *The Journal of Double Star Observations*.

For more information:
 The Journal of Double Star Observations
 http://www.southalabama.edu/physics/jdso/

METEOR OBSERVING

There are a number of organizations (note the previously listed Association of Lunar and Planetary Observers) specifically inclined to meteor watching, both naked eye and telescopic/binoculars.

Dr. Charles P. Olivier founded the American Meteor Society (AMS) in 1911. The AMS supports beginning visual meteor observers by providing instruction, charts, forms, and other materials upon receipt of a modest dues payment.

The AMS provides several publications of interest:

- *Meteor Trails* is the AMS quarterly journal containing technical articles, observing reports and analysis, and information on upcoming meteor showers.
- *AMS Annual Report* is a compendium of reports and the activities of the AMS.
- Special bulletins covering particular topics are also occasionally published.

The dues are a modest $6.50 for students and those actively involved in observing, $8.00 for associates, and $10.00 for groups.

To contact the AMS:

Karl Simmons
AMS Treasurer
3859 Woodland Heights
Callahan, FL 32011
e-mail: KSAM32011@aol.com

Robert Lunsford
Operations Manager
161 Vance Street
Chula Vista, CA
e-mail: lunsford@amsmeteors.org
www.amsmeteors.org

North American Meteor Network Internet coordinator Lew Gramer administers an e-mail list for all meteor enthusiasts. The list is frequented by beginners seeking to learn more about meteors and meteor observing, by experienced amateurs from around the world, and even occasionally by professional researchers.

www.meteorobs.org

The International Meteor Organization (IMO) was founded in 1988. The IMO was created in response to an ever growing need for international cooperation of amateur meteor observations. The collection of meteor observations by several methods from all around the world ensures the comprehensive study of meteor showers and their relation to comets and interplanetary dust. Annual dues are $20, with members receiving the bimonthly journal WGN (short for "Werkgroepnieuws").

To contact the IMO:
> Robert Lunsford
> Operations Manager
> 161 Vance Street
> Chula Vista, CA
> e-mail: lunsford@amsmeteors.org
> www.imo.net

FOR THOSE READERS IN CANADA OR UNITED KINGDOM:
British Astronomical Association
Founded in 1890, the British Astronomical Association (BAA) has an international reputation for the quality of its observational and scientific work. The BAA has a number of observational sections, including the sun, moon, Mercury and Venus, Mars, asteroids and remote planets, Jupiter, Saturn, comets, meteors, aurora, variable stars, and deep-sky objects

To contact the BAA:
> British Astronomical Association
> The Assistant Secretary
> Burlington House
> Piccadilly
> London W1J 0DU
> e-mail: office@britastro.org
> www.britastro.org/main/

Royal Astronomical Society of Canada
The beginnings of The Royal Astronomical Society of Canada (RASC) go back to the middle of the nineteenth century. Today there are 27 RASC "Centres" across Canada with over 4,900 mem-

bers worldwide. Like the Astronomical League, the RASC has several observing programs with others in development. RASC members receive an annual Observer's Handbook, journals, and access to a variety of guides and materials.

To contact the RASC:
    The Royal Astronomical Society of Canada
    136 Dupont Street
    Toronto, ON M5R 1V2
    Canada
    e-mail: nationaloffice@rasc.ca
    www.rasc.ca

Appendix D

Object Logs

OBJECT LOG INFORMATION

The circles on the object log provide you space to prepare up to two separate drawings of a specific object. These could be on different nights, with two different binocular magnifications, or just a rough, outdoor, at-the-binoculars sketch and an indoors refined sketch.

As noted below the data table, you should include at least two cardinal points (N-E-S-W) on the drawing as a reference.

Object identification can be its common name (i.e. Great Orion Nebula), catalog number (M42), or both (M42, Great Orion Nebula).

When recording date and time, astronomers usually use Universal Time, UT (sometimes referred to as Greenwich Mean Time or GMT, or Coordinated Universal Time or UTC). A 24-hour clock is also most often used. AAVSO variable star observations utilize the Julian Date (JD). You will need to convert from your local time to UT.

Seeing refers to the steadiness of the earth's atmosphere at your location. Are objects steady and not "wiggling" around or are they dancing all over the field of view? One way this is determined is by the maximum power you can use. Seeing is recorded as a number, usually from 1 to 10 or 1 to 5, with the highest number representing an outstanding night.

Transparency refers to how clear the skies are at your location. Transparency is also recorded as a number, usually from 1 to 10 or

0 to 7, with the highest number representing a completely clear and haze-free night.

Record the type of binoculars you are using, including the manufacturer (e.g., 7×50 Bushnell).

Record the place where you made your observations. Be as specific as possible; some record addresses whereas others record latitude and longitude.

Add additional comments as appropriate; this could include details you could not draw, something which impressed you, and so on.

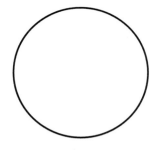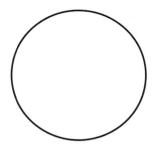

Object:

Date:	Time:
Seeing: (1–10 Scale; 10=best)	Transparency: (1–10 Scale; 10=best)
Binoculars:	Place:

Comments:

Indicate cardinal points on the drawing(s) above

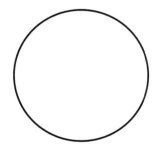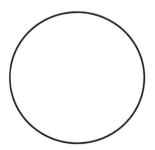

Object:

Date:	Time:
Seeing: (1–10 Scale; 10=best)	Transparency: (1–10 Scale; 10=best)
Binoculars:	Place:

Comments:

Indicate cardinal points on the drawing(s) above

OBJECT LOG TABLE

Some observers prefer an object log table instead of a form with circles to draw objects. A sample table is shown below.

Object	Date	Time	Seeing	Transparency	Binoculars	Place

Comments:

Object	Date	Time	Seeing	Transparency	Binoculars	Place

Comments:

Object	Date	Time	Seeing	Transparency	Binoculars	Place

Comments:

Appendix E

Lunar Features

Even with a pair of 7×50s, the moon is an inspiring object. The tables below, taken primarily from the Astronomical League's Lunar Observing Program and modified for binocular observers, lists features that are visible through binoculars. The tables are sequenced by lunar age. Features are arranged in each table from the moon's North Pole to the south.

The moon goes through its phases—from new moon to first quarter to full moon to third quarter back to new—in 29½ days. Features will be visible at different times, depending on their location. Those closest to the terminator will be the most obvious and interesting.

You can contact the Astronomical League website for more information on their Lunar Observing Program and Award Certificate.

APPROXIMATELY FOUR-DAY-OLD MOON

Feature	Crater	Mare	Mountains
Endymion	✓		
Atlas	✓		
Hercules	✓		
Cleomedes	✓		
Macrobius	✓		
Crisium		✓	
Proclus	✓		
Palus Somnii		✓	
Tranquillitatis		✓	
Fecunditatis		✓	
Langrenus	✓		
Vendelinus	✓		
Petavius	✓		

APPROXIMATELY SEVEN-DAY-OLD MOON
FIRST QUARTER

Feature	Crater	Mare	Mountains
Frigoris		✓	
Aristoteles	✓		
Eudoxus	✓		
Cassini	✓		
Lacus Somniorum		✓	
Aristillus	✓		
Posidonius	✓		
Autolycus	✓		
Serenitatis		✓	
Vaporum		✓	
Tranquillitatis		✓	
Sinus Medii		✓	

APPROXIMATELY SEVEN-DAY-OLD MOON
FIRST QUARTER (continued)

Feature	Crater	Mare	Mountains
Hipparchus	✓		
Albategnius	✓		
Theophilus	✓		
Cyrillus	✓		
Nectaris		✓	
Catharina	✓		
Altari Scarp			✓
Fracastorius	✓		
Piccolomini	✓		
Maurolycus	✓		

APPROXIMATELY TEN-DAY-OLD MOON

Feature	Crater	Mare	Mountains
Frigoris		✓	
Plato	✓		
Sinus Iridum		✓	
Archimedes	✓		
Apennine			✓
Imbrium		✓	
Aristarchus	✓		
Eratosthenes	✓		
Copernicus	✓		
Oceanus Procellarum		✓	
Ptolemaeus	✓		
Alphonsus	✓		
Gassendi	✓		
Arzachel	✓		
Nubium		✓	

APPROXIMATELY TEN-DAY-OLD MOON *(continued)*

Feature	Crater	Mare	Mountains
Humorum		✓	
Bullialdus	✓		
Palus Epidemiarum		✓	
Walter	✓		
Tycho	✓		
Maginus	✓		
Longomontanus	✓		
Clavius	✓		

APPROXIMATELY FOURTEEN-DAY-OLD MOON
FULL MOON

Feature	Crater	Mare	Mountains
Frigoris		✓	
Sinus Iridum		✓	
Sinus Roris		✓	
Kepler	✓		
Grimaldi	✓		
Oceanus Procellarum		✓	
Humorum		✓	
Nubium		✓	
Tycho's Crater Rays	✓& rays		

Appendix F

The Planets as Seen from Earth

Planet	Angular Equatorial Size		Magnitudes	
	Maximum	Minimum	Maximum	Minimum
Mercury	10	4.9	–1.3	
Venus	64	10	–4.4	
Mars	25.16	3.5	–2.9	–1.0
Jupiter	50.11	30.47	–2.9	–2.0
Saturn	20.75	18.44	–0.3	0.9
Saturn's Rings	46.7	41.5		
Uranus	3.96	3.6	5.65	6.06
Neptune	2.52	2.49	7.66	7.70
Pluto	0.11	0.065	13.6	15.95

Notes
- Angular equatorial size is in arc seconds
- Magnitudes for the inferior planets—Mercury and Venus—are given as maximum magnitudes only; this is the greatest brilliance that occurs, close to when the planet is the farthest from the sun as seen from Earth.
- Magnitudes for the superior planets—Mars, Jupiter, Saturn, Uranus, Neptune, and Pluto—are given as maximum and minimum magnitudes at opposition (when the planet is 180° opposite the sun as seen from Earth). This magnitude can vary depending on the positions of Earth's and the other planet's orbits when opposition occurs. Since the orbits are elliptical, the distances vary at opposition, and so does the magnitude as seen from Earth.
- Saturn's rings are measured at the widest part; magnitudes at opposition are the rings and planet combined.

Appendix G

The Messier Objects

Object	NGC	Constellation	Type	Magnitude	Size Arc Minutes	RA Hour Min	DEC Deg Min
M1	1952	Tau	SR	8.0	6×4	05 34.5	+22 01
M2	7089	Aqr	GC	6.3	12.9	21 33.5	−00 49
M3	5272	CVn	GC	5.9	16.2	13 42.2	+28 23
M4	6121	Sco	GC	5.4	26.3	16 23.6	−26 32
M5	5904	Ser	GC	5.7	17.4	15 18.6	+02 05
M6	6405	Sco	OC	4.2	15	17 40.1	−32 13
M7	6475	Sco	OC	2.8	80	17 53.9	−34 49
M8	6523	Sgr	OC/BN	3.0	60×35	18 03.8	−24 23
M9	6333	Oph	GC	7.8	9.3	17 19.2	−18 31
M10	6254	Oph	GC	6.6	15.1	16 57.1	−04 06
M11	6705	Sct	OC	5.3	14	18 51.1	−06 16
M12	6218	Oph	GC	6.8	14.5	16 47.2	−01 57
M13	6205	Her	GC	5.3	16.6	16 41.7	+36 28
M14	6402	Oph	GC	7.6	11.7	17 37.6	−03 15
M15	7078	Peg	GC	6.0	12.3	21 30.0	+12 10
M16	6611	Ser	OC/BN	6.0	7	18 18.8	−13 47
M17	6618	Sgr	BN	6.0	47×36	18 20.8	−16 11
M18	6613	Sgr	OC	6.9	9	18 19.9	−17 08
M19	6273	Oph	GC	6.8	13.5	17 02.6	−26 16

Object	NGC	Constellation	Type	Magnitude	Size Arc Minutes	RA Hour Min	DEC Deg Min
M20	6514	Sgr	BN	6.3	29×27	18 02.6	–23 02
M21	6531	Sgr	OC	5.9	13	18 04.6	–22 30
M22	6656	Sgr	GC	5.2	24.0	18 36.4	–23 54
M23	6494	Sgr	OC	5.5	27	17 56.8	–19 01
M24	6603	Sgr	MW	2.5	90	18 16.9	–18 29
M25	I4725	Sgr	OC	4.6	40	18 31.6	–19 15
M26	6694	Sct	OC	8.0	15	18 45.2	–09 24
M27	6853	Vul	PN	7.3	8.0×5.7	19 59.6	+22 43
M28	6626	Sgr	GC	6.9	11.2	18 24.5	–24 52
M29	6913	Cyg	OC	6.6	7	20 23.9	+38 32
M30	7099	Cap	GC	6.9	11.0	21 40.4	–23 11
M31	224	And	SG	3.4	178×63	00 42.7	+41 16
M32	221	And	EG	8.2	8×6	00 42.7	+40 52
M33	598	Tri	SG	5.7	73×45	01 33.9	+30 39
M34	1039	Per	OC	5.2	35	02 42.0	+42 47
M35	2168	Gem	OC	5.1	28	06 08.9	+24 20
M36	1960	Aur	OC	6.0	12	05 36.1	+34 08
M37	2099	Aur	OC	5.6	24	05 52.4	+32 33
M38	1922	Aur	OC	7.4	21	05 28.4	+35 50
M39	7092	Cyg	OC	4.6	32	21 32.2	+48 26
M40	Win4	UMa	DS	9.0, 9.6	0.8	12 22.4	+58 05

Note: This faint double star was found by Messier and was reported in error by Johann Hevelius in the seventeenth century. Messier noted no nebulosity around the stars, yet a catalog number was assigned.

Object	NGC	Constellation	Type	Magnitude	Size Arc Minutes	RA Hour Min	DEC Deg Min
M41	2287	CMa	OC	4.5	38	06 46.0	–20 44
M42	1976	Ori	BN	3.7	85×60	05 35.4	–05 27
M43	1982	Ori	BN	6.8	20×15	05 35.6	–05 16
M44	2632	Cnc	OC	3.1	95	08 40.1	+19 59
M45	–	Tau	OC	1.5	110	03 47.0	+24 07
M46	2437	Pup	OC	6.1	27	07 41.8	–14 49
M47	2422	Pup	OC	5.7	30	07 36.6	–14 30

Object	NGC	Constellation	Type	Magnitude	Size Arc Minutes	RA Hour Min	DEC Deg Min
M48	2548	Hya	OC	5.8	54	08 13.8	−05 48
M49	4472	Vir	EG	8.4	9×7.5	12 29.8	+08 00
M50	2323	Mon	OC	5.9	16	07 03.2	−08 20
M51	5194	CVn	SG	8.4	10×6	13 29.9	+47 12
M52	7654	Cas	OC	6.9	13	23 24.2	+61 35
M53	5024	Com	GC	7.7	12.6	13 12.9	+18 10
M54	6715	Sgr	GC	7.2	9.1	18 55.1	−30 29
M55	6809	Sgr	GC	6.3	19.0	19 40.0	−30 58
M56	6779	Lyr	GC	8.8	7.1	19 16.6	+30 11
M57	6720	Lyr	PN	8.8	1.4×1.0	18 53.6	+33 02
M58	4579	Vir	SG	9.7	5.5×4.5	12 37.7	+11 49
M59	4621	Vir	EG	9.6	5×3.5	12 42.0	+11 39
M60	4649	Vir	EG	8.8	7×6	12 43.7	+11 33
M61	4303	Vir	SG	9.6	6×5.5	12 21.9	+04 28
M62	6266	Oph	GC	6.7	14.1	17 01.2	−30 07
M63	5055	CVn	SG	8.6	10×6	13 15.8	+42 02
M64	4826	Com	SG	8.5	9.3×5.4	12 56.7	+21 41
M65	3623	Leo	SG	8.8	8×1.5	11 18.9	+13 05
M66	3627	Leo	SG	9.0	8×2.5	11 20.2	+12 59
M67	2682	Cnc	OC	6.0	15	08 50.4	+11 49
M68	4590	Hya	GC	7.6	12.0	12 39.5	−26 45
M69	6637	Sgr	GC	7.4	7.1	18 31.4	−32 21
M70	6681	Sgr	GC	7.8	7.8	18 43.2	−32 18
M71	6838	Sge	GC	8.0	7.2	19 53.8	+18 47
M72	6981	Aqr	GC	9.2	5.9	20 53.5	−12 32
M73	6994	Aqr	AS	8.9	2.8	20 58.9	−12 38
M74	628	Psc	SG	8.5	10.2×9.5	01 36.7	+15 47
M75	6864	Sgr	GC	8.6	6.0	20 06.1	−21 55
M76	650	Per	PN	10.1	2.7×1.8	01 42.4	+51 34
M77	1068	Cet	SG	8.9	7×6	02 42.7	−00 01

Object	NGC	Constellation	Type	Magnitude	Size Arc Minutes	RA Hour Min	DEC Deg Min
M78	2068	Ori	BN	8.0	8×6	05 46.7	+00 03
M79	1904	Lep	GC	7.7	8.7	05 24.5	−24 33
M80	6093	Sco	GC	7.3	8.9	16 17.0	−22 59
M81	3031	UMa	SG	6.9	21×10	09 55.6	+69 04
M82	3034	UMa	IG	8.4	9×4	09.55.8	+69 41
M83	5236	Hya	SG	7.5	11×10	13 37.0	−29 52
M84	4374	Vir	LG	9.1	5.0	12 25.1	+12 53
M85	4382	Com	LG	9.1	7.1×5.2	12 25.4	+18 11
M86	4406	Vir	LG	8.9	7.5×5.5	12 26.2	+12 57
M87	4486	Vir	EG	8.6	7.0	12 30.8	+12 24
M88	4501	Com	SG	9.6	7×4	12 32.0	+14 25
M89	4552	Vir	EG	9.8	4.0	12 35.7	+12 33
M90	4569	Vir	SG	9.5	9.5×4.5	12 36.8	+13 10
M91	4548	Com	SG	10.1	5.4×4.4	12 35.4	+14 30
M92	6341	Her	GC	6.5	11.2	17 17.1	+43 08
M93	2447	Pup	OC	6.2	22	07 44.6	−23 52
M94	4736	CVn	SG	8.2	7×3	12 50.9	+41 07
M95	3351	Leo	SG	9.7	4.4×3.3	10 44.0	+11 42
M96	3368	Leo	SG	9.2	6×4	10 46.8	+11 49
M97	3587	UMa	PN	9.9	3.4×3.3	11 14.8	+55 01
M98	4192	Com	SG	10.1	9.5×3.2	12 13.8	+14 54
M99	4254	Com	SG	9.9	5.4×4.8	12 18.8	+14 25
M100	4321	Com	SG	9.3	7×6	12 22.9	+15 49
M101	5457	UMa	SG	7.9	22.0	14 03.2	+54 21
M102?	5866	Dra	LG	10.0	5.2×2.3	15 06.5	+55 46

Note: There is some debate over whether M102 was ever properly identified by Messier. Some believe it was a duplication of M101, others that it is NCG 5866. This table assumes the latter.

Object	NGC	Constellation	Type	Magnitude	Size Arc Minutes	RA Hour Min	DEC Deg Min
M103	581	Cas	OC	7.4	6	01 33.2	+60 42
M104	4594	Vir	SG	8.0	9×4	12 40.0	−11 37
M105	3379	Leo	EG	9.3	2.0	10 47.8	+12 35

Object	NGC	Constellation	Type	Magnitude	Size Arc Minutes	RA Hour Min	DEC Deg Min
M106	4258	CVn	SG	8.3	19×8	12 19.0	+47 18
M107	6171	Oph	GC	7.8	10.0	16 32.5	−13 03
M108	3556	UMa	SG	10.0	8×1	11 11.5	+55 40
M109	3992	UMa	SG	9.8	7×4	11 57.6	+53 23
M110	205	And	EG	8.0	17×10	00 40.4	+41 41

KEY

AS	System of 4 Stars or Asterism	IG	Irregular Galaxy
BN	Bright Nebula	LG	Lenticular Galaxy
DN	Dark Nebula	SG	Spiral Galaxy
DS	Double Star	CW	Milky Way Patch
GC	Globular Cluster	PN	Planetary Nebula
OC	Open Cluster	SR	Supernova Remnant
EG	Elliptical Galaxy		

Appendix H

The Caldwell Objects

The Caldwell Catalog was created by popular British astronomer Sir Patrick Caldwell Moore to include many of the spectacular objects omitted by Charles Messier in the Messier Catalog. Unlike the Messier Catalog, the Caldwell Catalog also includes many of the spectacular objects in the Southern Hemisphere.

Object	NGC	Constellation	Type	Magnitude	Size Arc Minutes	RA Hour Min	DEC Deg Min
C1	188	Cep	OC	8.1	14	00 44.4	+85 20
C2	40	Cep	PN	12.4	0.6	00 13.0	+72 32
C3	4236	Dra	SG	9.7	21×7	12 16.7	+69 28
C4	7023	Cep	BN	6.8	18×18	21 01.8	+68 12
C5	IC342	Cam	SG	9.2	18×17	03 46.8	+68 06
C6	6543	Dra	PN	8.1	0.3/5.8	17 58.6	+66 38
C7	2403	Cam	SG	8.4	18×10	07 36.9	+65 36
C8	559	Cas	OC	9.5	4	01 29.5	+63 18
C9	Sh2-155	Cep	BN	7.7	50×10	22 56.8	+62 37
C10	663	Cas	OC	7.1	16	01 46.0	+61 15
C11	7635	Cas	BN	7.0	15×18	23 20.7	+61 12
C12	6946	Cep	SG	8.9	11×9	20 34.8	+60 09
C13	457	Cas	OC	6.4	13	01 19.1	+58 20
C14	869/884	Per	OC	4.3	30 & 30	02 20.0	+57 08

Object	NGC	Constellation	Type	Magnitude	Size Arc Minutes	RA Hour Min	DEC Deg Min
C15	6826	Cyg	PN	8.8	0.5/2.3	19 44.8	+50 31
C16	7243	Lac	OC	6.4	21	22 15.3	+49 53
C17	147	Cas	EG	9.3	13×8	00 33.2	+48 30
C18	185	Cas	EG	9.2	12×9	00 39.0	+48 20
C19	IC5146	Cyg	BN	10.0	12×12	21 53.5	+47 16
C20	7000	Cyg	DN	6.0	120×100	20 58.8	+44 20
C21	4449	CVn	IG	9.4	5×3	12 28.2	+44 06
C22	7662	And	PN	8.3	0.3/2.2	23 25.9	+42 33
C23	891	And	SG	9.9	14×2	02 22.6	+42 21
C24	1275	Per	IG	11.6	2.6×1	03 19.8	+41 31
C25	2419	Lyn	GC	10.4	4.1	07 38.1	+38 53
C26	4244	CVn	SG	10.2	16×2.5	12 17.5	+37 49
C27	6888	Cyg	BN	7.5	20×10	20 12.0	+38 21
C28	752	And	OC	5.7	50	01 57.8	+37 41
C29	5005	CVn	SG	9.8	5.4×2	13 10.9	+37 03
C30	7331	Peg	SG	9.5	11×4	22 37.1	+34 25
C31	IC405	Aur	BN	6.0	30×19	05 16.2	+34 16
C32	4631	CVn	SG	9.3	15×3	12 42.1	+32 32
C33	6992/5	Cyg	SN		60×8	20 56.4	+31 43
C34	6960	Cyg	SN		70×6	20 45.7	+30 43
C35	4889	Com	EG	11.4	3×2	13 00.1	+27 59
C36	4559	Com	SG	9.8	10×4	12 36.0	+27 58
C37	6885	Vul	OC	5.9	7	20 12.0	+26 29
C38	4565	Com	SG	9.6	16×3	12 36.3	+25 59
C39	2932	Gem	PN	9.2	0.2/0.7	07 29.2	+20 55
C40	3626	Leo	SG	10.9	3×2	11 20.1	+18 21
C41	–	Tau	OC	0.5	330	04 27.0	+16 00
C42	7006	Del	GC	10.6	2.8	21 01.5	+16 11
C43	7814	Peg	SG	10.5	6×2	00 03.3	+16 09
C44	7479	Peg	SG	11.0	4×3	23 04.9	+12 19

Object	NGC	Constellation	Type	Magnitude	Size Arc Minutes	RA Hour Min	DEC Deg Min
C45	5248	Boo	SG	10.2	6×4	13 37.5	+08 53
C46	2261	Mon	BN	10.0	2×1	06 39.2	+08 44
C47	6934	Del	GC	8.9	5.9	20 34.2	+07 24
C48	2775	Cnc	SG	10.3	4.5×3	09 10.3	+07 02
C49	2237-9	Mon	BN		80×60	06 32.3	+05 03
C50	2244	Mon	OC	4.8	24	06 32.4	+04 52
C51	IC1613	Cet	IG	9.3	12×11	01 04.8	+02 07
C52	4697	Vir	EG	9.3	6×3	12 48.6	−05 48
C53	3115	Sex	EG	9.1	8×3	10 05.2	−07 43
C54	2506	Mon	OC	7.6	7	08 00.2	−10 47
C55	7009	Aqr	PN	8.3	2.5/1	21 04.2	−11 22
C56	246	Cet	PN	10.9	3.8	00 47.0	−11 53
C57	6822	Sgr	IG	8.8	10×9	19 44.9	−14 48
C58	2360	CMa	OC	7.2	13	07 17.8	−15 37
C59	3242	Hya	PN	7.8	0.3/21	10 24.8	−18 38
C60	4038	CrV	SG	10.7	2.6×1.8	12 01.9	−18 52
C61	4039	CrV	SG	10.7	3.2×2.2	12 01.9	−18 53
C62	247	Cet	SG	9.1	20×7	00 47.1	−20 46
C63	7293	Aqr	PN	7.3	13	22 29.6	−20 48
C64	2362	CMa	OC	4.1	8	07 18.8	−24 57
C65	253	Scl	SG	7.1	25×7	00 47.6	−25 17
C66	5694	Hya	GC	10.2	3.6	14 39.6	−26 32
C67	1097	For	SG	9.2	9×6	02 46.3	−30 17
C68	6729	CrA	BN	9.7	1	19 01.9	−36 57
C69	6302	Sco	PN	9.6	0.8	17 13.7	−37 06
C70	300	Scl	SG	8.7	20×13	00 54.9	−37 41
C71	2477	Pup	OC	5.8	27	07 52.3	−38 33
C72	55	Scl	SG	7.9	32×6	00 14.9	−39 11
C73	1851	Col	GC	7.3	11	05 14.1	−40 03
C74	3132	Vel	PN	9.4	0.8	10 07.7	−40 26

Object	NGC	Constellation	Type	Magnitude	Size Arc Minutes	RA Hour Min	DEC Deg Min
C75	6124	Sco	OC	5.8	29	16 25.6	−40 40
C76	6231	Sco	OC	2.6	15	16 54.0	−41 48
C77	5128	Cen	EG	7.0	18×14	13 25.5	−43 01
C78	6541	CrA	GC	6.6	13	18 08.0	−43 42
C79	3201	Vel	GC	6.7	18	10 17.6	−46 25
C80	5139	Cen	GC	3.6	36	13 26.8	−47 29
C81	6352	Ara	GC	8.1	7	17 25.5	−48 25
C82	6193	Ara	OC	5.2	15	16 41.3	−48 46
C83	4945	Cen	SG	8.7	20×4	13 05.4	−49 28
C84	5286	Cen	GC	7.6	9	13 46.4	−51 22
C85	IC2391	Vel	OC	2.5	50	08 40.2	−53 04
C86	6397	Ara	GC	5.6	26	17 40.7	−53 40
C87	1261	Hor	GC	8.4	7	03 12.3	−55 13
C88	5823	Cir	OC	7.9	10	15 05.7	−55 36
C89	6087	Nor	OC	5.4	12	16 18.9	−57 54
C90	2867	Car	PN	9.7	0.2	09 21.4	−58 19
C91	3532	Car	OC	3.0	55	11 06.4	−58 40
C92	3372	Car	BN	3	120×120	10 43.8	−59 52
C93	6752	Pav	GC	5.4	20	19 10.9	−59 59
C94	4755	Cru	OC	4.2	10	12 53.6	−60 20
C95	6025	TrA	OC	5.1	12	16 03.7	−60 30
C96	2516	Car	OC	3.8	30	07 58.3	−60 52
C97	3766	Cen	OC	5.3	12	11 36.1	−61 37
C98	4609	Cru	OC	6.9	5	12 42.3	−62 58
C99	–	Cru	DN		400×300	12 53.0	−63 00
C100	IC2944	Cen	OC	4.5	15	11 36.6	−63 02
C101	6744	Pav	SG	8.3	16×10	19 09.8	−63 51
C102	IC2602	Car	OC	1.9	50	10 43.2	−64 24
C103	2070	Dor	BN	3	40×25	05 38.7	−69 06
C104	362	Tuc	GC	6.6	13	01 03.2	−70 51

Object	NGC	Constellation	Type	Magnitude	Size Arc Minutes	RA Hour Min	DEC Deg Min
C105	4833	Mus	GC	7.3	14	12 59.6	−70 53
C106	104	Tuc	GC	4	31	00 24.1	−72 05
C107	6101	Aps	GC	9.3	11	16 25.8	−72 12
C108	4372	Mus	GC	7.8	19	12 25.8	−72 40
C109	3195	Cha	PN	11.6	0.6	10 09.5	−80 52

KEY

AS	System of 4 Stars or Asterism	IG	Irregular Galaxy
BN	Bright Nebula	LG	Lenticular Galaxy
DN	Dark Nebula	SG	Spiral Galaxy
DS	Double Star	CW	Milky Way Patch
GC	Globular Cluster	PN	Planetary Nebula
OC	Open Cluster	SR	Supernova Remnant
EG	Elliptical Galaxy		

SIZE

Object sizes are given in arc minutes. For many of the planetary nebula, the size is given in the form X/Y, where X represents the diameter and Y the planetary nebula's halo.

Appendix I

Planispheres

For beginners and many seasoned amateurs alike, the planisphere provides an easy way to find your way around the night sky. Basically the planisphere is simply a circular star map which rotates to show the stars and constellations visible at any date and time during the year. The planisphere is an excellent way to learn constellations and stars, as well as point you in the right direction while using your binoculars in locating objects.

Setting a planisphere is simple. You will note that a star wheel turns inside of an exterior overlay; the star wheel turns on a grommet which holds it and the overlay together. Along the exterior edge of the overlay and the star wheel you will find the date and the time. Turn the wheel to match the time you wish to observe to the date; it is that simple to set! This roughly represents the sky as you would see it on the date and time you have set.

To use the planisphere may take a little getting used to the first couple of times, so be patient. Remember that you will need a red-filtered flashlight; you do not want to shine a white light on the planisphere then try to find stars and constellations in the night sky.

I hold the planisphere in front of me, just above my head; this allows me to glace at the planisphere then up to the sky. The planisphere's orientation is important; turn it so that the edge of the planisphere labeled with the direction you are facing is at the bottom, away from you. For example, if you are facing south, the south portion of the planisphere will be at the bottom away from you and north at the top, closest to you. East will be on your left, west to the right.

You might quickly note some of the problems with planispheres. They are relatively small and distorted (because the planisphere translates a 3-dimensional hemisphere onto a flat surface). Some try to get around the size problems by larger planispheres, such as David Levy's large 16-inch diameter planisphere. Distortion is dealt with fairly effectively by David Chandler's The Night Sky Planisphere and Sky & Telescope's Star Wheel by showing the sky on two sides, north and south views.

When selecting a planisphere, look at it to see if you can easily read the information on the planisphere itself. I personally do not find glow-in-the-dark paint to be effective. Also look to see if the planisphere differentiates between star brightnesses by recognizable different size dots; this will help you in properly identifying stars. The final selection criterion is to note your latitude; planispheres are usually sold by a latitude range such as 20 degrees to 30 degrees N.

A couple of closing notes. Planispheres do not show the location of the moon and planets since they are constantly moving against the background stars. If you are using your planisphere during daylight savings time, you must subtract an hour. Also recognize that the planisphere—and your clock—may say 10 P.M., but the sky might not. This is referred to as local mean time and it is how the sky actually appears where you live. The scale of a planisphere is such that adjusting for local mean time usually is not done by many observers.

On the following pages are sources for planispheres, including those you can download and make yourself, commercial planispheres, planispheres for PDAs, and additional directions. This is not an exhaustive list of sources and resources; however, it will provide you with a starting point.

MAKE-YOUR-OWN PLANISPHERES
M. S. Pettersen, Washington and Jefferson College:
    http://www.washjeff.edu/physics/plan.html
Jan Tosovsky, NIO's computer planisphere:
    http://nio.astronomy.cz/om/

Brian Tung: http://astro.isi.edu/games/planisphere.html
Ken Wilson, Science Museum of Virginia CD Planisphere:
  http://www.smv.org/pubs/cdplanispherecomplete2.doc

Note: There are a number of "make it yourself" planispheres on-line that will provide you will a satisfactory planisphere and possibly a better understanding of how the planisphere works.

READY-MADE PLANISPHERES

Davd Chandler Company. The Night Sky Planisphere. Sizes: Small (5 inch by 6 inch) and Large (8½ inch by 10 inch), Northern Hemisphere latitude ranges: 20°–30°, 30°–40°, 40°–50°, 50°–60°, Southern Hemisphere: 30°–40° South, Japanese: 35° N. Post Office Box 999, Springville, CA 93265.

Storm Dunlop and Wil Tirion. The Firefly Planisphere Deluxe. A heavy-duty 15-inch diameter planisphere for Northern Hemisphere latitude ranges of 40°–60°.

David Kennedal. Precision Planet and Star Locator. An 11-inch diameter planisphere for Northern Hemisphere latitude ranges of 30°, 35°, 50°, and 30° to 60°.

David Levy. David H. Levy's Guide to the Stars. A large 16-inch diameter planisphere (largest size available) for Northern Hemisphere latitude ranges of 30°–60°.

Miller Planisphere. Pocket (5½ inch) and Classic (10½ inch) sizes for Northern Hemisphere latitude ranges: 17°–27°, 25°–35°, 35°–45°, 45°–55°.

George Philips. The Philips Planisphere.

Sky & Telescope's Star Wheel. A 10¾ inch by 11½ inch planisphere for Northern Hemisphere latitude ranges of 30°, 40°, 50°, and 30° South Latitude.

PLANISPHERE DIRECTIONS

Discovery Channel's School:
  http://school.discovery.com/schooladventures/skywatch/
    howto/planisphere1.html
Alan M. MacRobert, Sky & Telescope:
  http://www.astro-tom.com/getting_started/using_a_
    planisphere.htm
  http://skyandtelescope.com/howto/visualobserving/
    article_75_1.asp

Stig's Sky Calendar:
    http://www.skycalendar.com/skycal/help/calinfo6.html
Planispheres for PDAs:
    http://members.aol.com/psphere/
Planisphere 4.31:
    http://palmsource.palmgear.com/index.cfm?fuseaction=
    software.showsoftware&prodid=51062

WITHDRAWN

*Also available from Mike D. Reynolds
and Stackpole Books*

FALLING STARS

A straightforward and practical guide
to meteors and meteorites

*$14.95, paperback, 160 pages, 54 photographs,
0-8117-2755-6*

Available from your favorite bookstore or

STACKPOLE BOOKS
800-796-0411
www.stackpolebooks.com